The Sun, the Genome, & the Internet

Each year The New York Public Library and Oxford University Press invite a prominent figure in the arts and letters to give a series of lectures on a topic of his or her choice. The lectures become the basis of a book jointly published by the Library and the Press. The previous books in the series are *The Old World's New World* by C. Vann Woodward; *Culture of Complaint: The Fraying of America* by Robert Hughes; *Witches and Jesuits: Shakespeare's Macbeth* by Garry Wills; *Visions of the Future: The Distant Past, Yesterday, Today, Tomorrow* by Robert Heilbroner; and *Doing Documentary Work* by Robert Coles.

The sun, the genome, the internet

Tools of Scientific Revolutions

Freeman J. Dyson

The New York Public Library
Oxford University Press
New York   Oxford
1999

Oxford University Press

Oxford   New York
Athens   Auckland   Bangkok   Bogotá   Buenos Aires   Calcutta
Cape Town   Chennai   Dar es Salaam   Delhi   Florence   Hong Kong   Istanbul
Karachi   Kuala Lumpur   Madrid   Melbourne   Mexico City   Mumbai
Nairobi   Paris   São Paulo   Singapore   Taipei   Tokyo   Toronto   Warsaw

and associated companies in
Berlin   Ibadan

Published by Oxford University Press, Inc.
198 Madison Avenue, New York, New York 10016

Oxford is a registered trademark of Oxford University Press

Library of Congress Cataloging-in-Publication Data
Dyson, Freeman J.
The sun, the genome, and the internet: tools of scientific revolutions / Freeman J. Dyson
p. cm. Includes bibliographical references.
ISBN 0-19-512942-3 (alk. paper)
1. Mathematical physics.   2. Science—History.   3. Science—Philosophy
I. Title
QC20.D98   1999
303.48'3—dc21   98-53830

9 8 7 6 5 4 3 2 1
Printed in the United States of America
on acid-free paper

Contents

introduction

When I was seventeen years old I came as a student to Cambridge University and got to know the famous mathematician Godfrey Hardy.

Almost all the mathematicians and scientists were away, helping to fight the Second World War. There were no graduate students and very few advanced students of any kind. All that was left of the mathematical life of Cambridge was the old and famous professors and a handful of very young undergraduates. Hardy, who was then sixty-four years old, was depressed and miserable. He was already suffering from the heart ailment that would cripple him a few years later. He never spoke about the war that was raging around us; he abominated war so deeply that he could not bring himself to speak about it. He gave lectures on the pure mathematics that he loved to four or five students sitting around a small table in a small seminar room. In that little room we sat within a couple of feet of Hardy, three times a week for two years. He lectured like Wanda Landowska playing Bach on the harpsichord: precise and totally lucid, but displaying his passionate pleasure to all who could see beneath the surface. Each lecture was carefully prepared, like a work of art, with the intellectual dénouement appearing as if spontaneously in the last five minutes of the hour. For me these lectures were an intoxicating joy, and I used to feel sometimes an impulse to hug that little old man with the white hair two feet away from me, to show him how desperately grateful we were to him for his willingness to go on talking.

A year before I came to Cambridge, Hardy had published a little book with the title *A Mathematician's Apology*. The book was written for readers with no mathematical training, introducing them gently into the world of mathematics in which he was at home. He was a very pure mathematician, and the message of his book is that pure mathematics is the only kind of mathematics worthy of respect. He wrote: "A mathematician, like a painter or a poet, is a maker of patterns. If his patterns are more permanent than theirs, it is because they are made with ideas." He saw himself as an artist, creating works of abstract beauty. He saw applied mathematics as second-rate mathematics, more often doing harm than good, and hated applied mathematics with special intensity when it had anything to do with war. He proudly claimed that he had never done anything in his life that could be considered useful. Everything he did was a work of art and was done with style. His mathematical papers are beautiful in style as well as in content. He wrote as clearly and as elegantly as he thought.

In my later scientific activities I was unfaithful to

Hardy's dream. I began by following his footsteps into number theory and solved a couple of number-theoretical problems. These problems were elegant but unimportant. Then, after three years as a number theorist, I became an applied mathematician. I decided it would be more exciting to understand one of the basic mysteries of nature than to continue proving theorems that were of interest only to a small coterie of number theorists. I carefully avoided speaking to Hardy when I made this decision, because I knew it would only cause him grief. A few years later, when I had successfully made the transition to applied mathematics and felt free to talk to him about it, he was already dead. I wanted to tell him that there was no difference in quality between what I had done as a number theorist and what I was doing as a physicist. But by then it was too late.

Many times in later years I have wished that I could explain to Hardy what I have done with the mathematics that he taught me. I dream sometimes that he will understand and forgive me for straying from his ideals.

Throughout my professional life, I have been happily finding areas of science where my mathematical skills could be usefully employed. I have worked on a variety of problems in particle physics, in statistical mechanics, in condensed-matter physics, in astronomy, and in biology. I also have worked on engineering problems, applying mathematics to the design of instruments and machines. When I was designing machines, I often thought of the famous statement in Hardy's book, the statement that expressed in a few bitter words his hatred of applied science: "A science is said to be useful if its development tends to accentuate the existing inequalities in the distribution of wealth, or more directly promotes the destruction of human life." I was trying to prove Hardy wrong, trying to prove that science can be useful without being harmful. In choosing problems to work on, I always kept Hardy's warning in mind. Hardy's statement is often true, and it is a warning that all applied scientists must take seriously.

Three times in my life, I worked on applied projects that Hardy might have found objectionable. In 1956 I helped to design an inherently safe nuclear reactor, which was sold with the name TRIGA and is still being manufactured and sold today. It is mostly used in hospitals to produce short-lived isotopes for diagnosis of metabolic diseases. Short-lived isotopes have many advantages, but they must be produced in the same place where they are used. The inherent safety of the reactor is based on the laws of physics and not on mechanical devices. The prototype TRIGA was demonstrated to the world at a public ceremony in San Diego, with the great physicist Niels Bohr himself pushing the switch to start it running. In those days the majority of scientists, like Bohr, thought of peaceful nuclear energy as a blessing to mankind. Although nuclear energy did not fulfill all of our hopes, I still believe that the TRIGA is a tool for good and not for evil.

Another engineering project that I worked on was adaptive optics. The problem was to design a flexible mirror with a rapid control system, to compensate the distortion of optical images produced by turbulence in the atmosphere. Adaptive optics can in principle allow a ground-based telescope to see objects in the sky as clearly as a space-based telescope. Unfortunately the same technology which helps peaceful astronomers to see clearly can also be used to focus laser beams for military purposes. Before I began my work, I made a careful study of the peaceful and the military applications of adaptive optics. I came to the conclusion that the death ray applications were fantasies while the astronomical applications were real. It now appears that my judgment was correct. Adaptive optics are being used by astronomers with great success, while the beam weapons are seen only in science fiction stories and films.

My latest involvement in a practical application of mathematics happened a couple of years ago. The problem is to help the local people in Cambodia and Bosnia and Angola and Afghanistan, the countries where millions of land mines are hidden in the ground and thousands of innocent civilians and children are killed and crippled every year. These people need mine detectors that are cheap and portable, cheap enough to be mass-produced and widely distributed, portable enough to be useful in places without roads and electricity. The U.S. Army is willing to help but has the wrong kind of equipment. The army's equipment is heavy and expensive, designed to clear a minefield fast so that soldiers can move through it rapidly. The problem is to adapt the army's technology to civilian needs by making it light and simple. For civilians, speed is not essential.

Paul Horowitz, a physicist at Harvard who is one of the leaders of the search for messages from extraterrestrial life-forms, is also a leader of efforts to solve the mine problem. He invented what he calls a smart probe, which is a modified version of the pointed stick normally used by people searching for mines in the ground. The normal method is to poke very carefully with the stick around a buried object, distinguishing mines from rocks and tree roots by their shape.

Horowitz designed a stick with a tiny acoustic transmitter and receiver hidden inside its point. The point transmits a succession of gentle sound-pulses into any object that it touches and receives a response delineating the internal vibrations of the object. Mines give a response that is quite different from the response of rocks and tree roots. The process of identifying mines becomes much quicker and safer. The response even tells you which kind of mine it is.

Mines manufactured in Russia, Italy, China, and the United States produce easily distinguishable responses. The smart probe uses very little electric power and can be run on batteries. Horowitz is now engaged in making it light and rugged enough to be used in the field. The aim is to make it as similar as possible to the existing stick probe, so that people who have grown accustomed to using the stick probe will easily learn to use the smart probe. I was happy to be working with Horowitz on the mine problem. In a small way, he is undoing the harm that the abuse of science has done to mankind. I believe that Hardy, if he knew what we were doing, would give us his blessing.

*   *   *

Technology is only one of many forces driving human history, and seldom the most important. Politics and religion, economics and ideology, military and cultural rivalries are at least as important as technology. Technology only gives us tools. Human desires and institutions decide how we use them. I emphasize technology in this book because that is where I come from. When I write about human desires and institutions, I write as a layman. When I write about technology, I am to some degree an expert.

Experts, when they try to forecast the future, are usually wrong.

In this book I am looking into the twenty-first century from the end of the twentieth. Before writing it, I examined in some detail the efforts of two famous experts who looked into the twentieth century from the end of the nineteenth, Jules Verne and H. G. Wells. Neither Verne nor Wells saw the future clearly. It is not fair to blame them for their mistakes and omissions, since they were storytellers rather than scientists. Their purpose was to entertain their readers and not to analyze technologies. Jules Verne, in his essay "An Ideal City," written in 1875, foresees a rising birthrate leading to the installation of a "five-hundred-nurse-power baby-feeding machine." Wells, in his novel *When the Sleeper Wakes*, written in 1899, foresees hypnotism replacing drugs, antiseptics, and anesthetics as the main working tool of medical practice. Neither of these fantasies was intended to be a serious prediction.

Verne's most successful predictions were Captain Nemo's submarine in *Twenty Thousand Leagues Under the Sea* and the Baltimore Gun Club's mission to the moon in *From the Earth to the Moon*. Wells scored more successes in *When the Sleeper Wakes*. He described a color television displaying remote events instantaneously, airplanes flying from London to America in two hours, and a CAD-CAM machine (computer-aided design, computer-aided manufacture) supplying the Sleeper quickly with new clothes after he woke up. Verne and Wells both failed to foresee some of the most basic aspects of twentieth-century life: the dominance of the private automobile over other means of transportation, the ubiquity of telephones and personal computers, the rapid disappearance of colonial empires. Their most basic mistake was their belief that technology would always become bigger and bigger rather than smaller and smaller. Wells had the population of England crammed into five cities, with thirty-three million people in London, the countryside totally deserted. Ed Tenner in his book *Why Things Bite Back* remarks that Albert Robida, a French cartoonist with no technical training and no love for technology, portrayed the twentieth century more accurately than either Verne or Wells. Robida in 1883 published cartoon pictures of flat television screens, test-tube babies, bomber airplanes, and chemical warfare. Robida understood, as Verne and Wells did not, that the twentieth would be the century of world wars.

Since Verne and Wells were unable to predict the

future accurately a hundred years ago, I do not expect that my vision today will be less full of errors than theirs. I am describing one possible course among the million other courses that the future might take. The probability that things will happen in the way that I describe is very small. I am concerned only to show that one possible road to a better future exists. Either along this road that I describe, or more probably along some other road, we have a chance to reach a happier and more equitable world. It is not important that we correctly identify the road before we reach it. The purpose of this book is to encourage us to search for it.

As a scientist I make a sharp distinction between models and theories. A theory is a construction, built out of logic and mathematics, that is supposed to describe the actual universe that we live in. A model is a construction that describes a much simpler universe, including some features of the actual universe and neglecting others. Theories and models are both useful tools for understanding nature. They are useful in different ways, and it is important to keep the difference in mind. A theory is useful because it can be tested by comparing its predictions with observations of the real world. On the other hand, a theory may be useless because its consequences are too complicated to be predicted. A model is useful because its behavior is simple enough to be predicted and understood. On the other hand, a model may be useless because it leaves out too much and loses any connection with reality. As we explore the universe, we move out from well-trodden ground into the unknown. On well-trodden ground we build theories. On the half-explored frontiers we build models. For example, in astronomy, Einstein's theory of general relativity gives a marvelously accurate picture of things that are close to us in space and time. When we move out to the remote frontiers, we describe the universe with cosmological models. The cosmological models have names like "New Inflationary Universe" or "Big Bang with Cold Dark Matter." Models are essential tools for under-ſtanding nature in regions beyond the reach of obser-vation. Later, when more details can be observed, models will be replaced by theories.

The hiſtory of technology in the twenty-firſt cen-tury is a prime example of territory beyond the reach of observation. My piƈture of it is a model and not a the-ory, omitting essential features of the real world. Accu-rate prediƈtion is not my purpose. The purpose of model-building in the realm of the future is not to prediƈt, but only to draw a rough sketch of the territo-ry into which we might be moving.

*   *   *

This book is based on a set of leƈtures given at the New York Public Library in the spring of 1997. The title of the series was "Three Faces of Science." The leƈtures consiſted moſtly of ſtories about science, with digressions into hiſtory and politics. Some of the ſtories were concerned with the paſt, others with the future. I was telling ſtories to show to an audience of nonscientiſts how science could be important in their lives. I am grateful to the New York Public Library audiences for helping me to identify the essential issues. In converting the leƈtures into a book, I have left the ſtories more or less as I told them. I have not tried to eliminate the digressions. I have only rearranged the material so as to emphasize the main theme, a model of the future in which the sun, the genome, and the internet are the driving forces.

I am grateful to the New York Public Library, and to Pamela Leo in particular, for organizing the leƈtures and providing me with lively audiences in a splendid leƈture hall. I am grateful to Oxford University Press, and to Laura Brown in particular, for financing the lectures and agreeing to publish the book. I am grateful to many other individuals for their help along the way, and especially to my daughter Esther and son George, whose books, *Release 2.0* and *Darwin Among the Machines*, have had a profound influence on my thinking. To my wife, Imme, I am grateful, as always, for listening to the lectures and for being my most honest critic.

The Sun, the Genome, & the Internet

I

scientific Revolutions

John Randall was in 1939 a thirty-four-year-old English physicist who had made an undistinguished career in solid-state physics. He was generally considered to be a hard worker but not very bright.

In September 1939 England was suddenly faced with a deadly enemy in the person of Adolf Hitler. England was miserably unprepared to defend itself against bombing or against air-supported invasion. The primitive radar system deployed around the English coast used meter-wave transmitters. These were then the only transmitters powerful enough to produce detectable echoes from an aircraft a hundred miles away. As a first line of defense for England they had three serious defects. First, they could only measure the distance of an aircraft and had very poor resolution in angle. Second, their effective range would be drastically reduced as soon as the Germans started to jam them with radio noise on the same frequency. And third, the worst defect of all, they were almost totally ineffective against low-flying aircraft. German intruders flying low over the sea would usually be able to reach the coast without detection.

Everybody who knew how bad the situation was also knew the cure for it. The cure was microwaves. Radars working with microwaves, with centimeter wavelengths, would be enormously more effective than meter-wave radars. They would have narrow search-light beams, giving them good discrimination in angle as well as in range. The narrow beam would also give them protection against jamming and good perfor-mance against low-flying aircraft. Unfortunately, nobody knew how to build a microwave transmitter strong enough to make a useful radar. The best trans-mitters in those days were putting out a few watts of power when kilowatts were needed. John Randall was then a junior assistant in the physics department of the University of Birmingham. He had been working on the physics of microwaves. In September 1939 he was given the job of inventing a high-power microwave transmitter. With the help of an engineer colleague, Henry Boot, Randall solved the problem in two months. He invented the cavity magnetron in Novem-ber 1939, and within eight months he had a working model putting out twelve kilowatts of microwaves, more than a thousand times the power of any earlier transmitter. The deadly enemy gave him the push to do something brilliant. Grim necessity multiplied his brainpower, just as his invention multiplied the trans-mitter power. A cavity magnetron was secretly brought from England to America in 1941, before America entered the war, as part of a bargain between Churchill and Roosevelt. It powered the microwave radars that gave a crucial advantage to British and American air forces and contributed heavily to Adolf Hitler's defeat. This is the first half of the story of John Randall.

The second half of the story begins when World War II ends.

Randall was now famous, his enormous contribution to victory acclaimed by a grateful nation. He was appointed full professor and head of department at King's College London, which in those days meant that he was the boss, with the power and prestige to do whatever he liked. He understood that he had a unique chance to be as creative in peace as he had been in war. He decided to abandon his old career in solid-state physics and microwaves and make a completely fresh start. He decided to create a department in the embryonic field of molecular biology. He quickly obtained substantial funding from two independent sources, the Medical Research Council in England and the Rockefeller Foundation in the United States. He spent a few years learning what was known about molecular biology and fixing up his heavily bombed building in London. He installed the best available equipment for X-ray crystallography and invited Maurice Wilkins and Rosalind Franklin to come and use it. Within a few more years, Wilkins and Franklin took the first clear pictures of the X-ray diffraction produced by aligned fibers of DNA. Their pictures led to the theoretical elucidation of the double helix structure of dna by Crick and Watson. It was directly due to Randall's vision that the structure of DNA was discovered in 1953 in England and not a few years later in the United States. Randall would probably never have dreamed his vision, and would certainly not have had the chance to bring it to reality, if he had not had greatness thrust upon him by Adolf Hitler.

Randall provided tools for three scientific revolutions. The great revolution in molecular biology that followed the discovery of the double helix was driven by the new tool of X-ray crystallography. Meanwhile, two more revolutions were occurring in atomic physics and radio astronomy. Immediately after World War II, the tools of microwave technology developed during the war for military purposes were applied to the spectroscopy of atoms and molecules and to the detection of microwave sources in the sky. Microwave spectroscopy revolutionized atomic physics by giving us a thousand-fold improvement in the sharpness of our view of the fine-structure of atoms. Microwave observations revolutionized astronomy by giving us pictures of radio sources in the sky with sharpness equal to the best pictures obtained from optical telescopes. These two revolutions were driven by technologies that grew out of Randall's invention of the cavity magnetron, while he was already moving on to provide the tools for the revolution in molecular biology.

The story of John Randall has many parallels in the generation of scientists who were young during World War II and had greatness thrust upon them by the exigencies of war. I think particularly of my friends Robert Wilson and Richard Feynman, who were rising stars in the crowd of young people who went to Los Alamos in 1943 to build nuclear bombs. Wilson rose to be head of experimental physics at Los Alamos, and Feynman was organizer of the numerical calculations of bomb performance. As soon as the war was over, both of them swore never again to have anything to do with nuclear bombs. Both of them switched from bombs to peaceful science and rapidly became international leaders. Wilson was the leading builder of particle accelerators for thirty years after the war, while Feynman became the leading particle theorist. It was largely due to the outstanding abilities of Wilson and Feynman that the United States became for thirty years the best place in the world to do particle physics.

Los Alamos gave them a vision of the power of science to shake the world. And Los Alamos gave them the confidence to be pioneers in the new world of particle physics, just as the cavity magnetron gave Randall the confidence to be a pioneer in molecular biology.

Fifty years later, the same theme of war and peace reappears in the world of medicine. Today the fight against AIDS is as bitter and as tragic as the fight against Hitler fifty years ago. AIDS is as deadly an enemy as Hitler. AIDS will not surrender unconditionally, even after major defeats. The fight against AIDS must be fought to a finish, just as the fight against Hitler was fought to a finish. Some day there will be a cure for AIDS at a price the victims can afford, and then the fight will be over. When it is over, the young doctors and virologists who helped win it will have greatness thrust upon them. They will be moving out from the medical battlefields into the wider world of science. The most enterprising of them will then have the prestige and the public support that will allow them to do whatever they like. I hope they will be ready when that time comes. I hope they will be ready to do something new, to launch new branches of science, to build new tools that will explore new worlds. If they are lucky, they will find themselves, like John Randall, leading new scientific revolutions. If they are very lucky, they will find themselves, like Randall, providing tools for three scientific revolutions simultaneously. The medical scientists of today will be leading revolutions into directions that we can only dimly discern.

*   *   *

Science originated from the fusion of two old traditions, the tradition of philosophical thinking that began in ancient Greece and the tradition of skilled crafts that began even earlier and flourished in medieval Europe. Philosophy supplied the concepts for science, and skilled crafts supplied the tools. Until the end of the nineteenth century, science and craft industries developed along separate paths. They frequently borrowed tools from each other, but each maintained an independent existence. It was only in the twentieth century that science and craft industries became inseparably linked.

My grandfather John William was a blacksmith, working in a small ironworks in Yorkshire, in the north of England. With his own hands he forged boilers that were exported all over the world, bringing steam power to drive ships and start industrial revolutions in remote places. He was not much concerned with science. He was a skilled craftsman, still working in the old tradition of craft industry that started the first industrial revolution in England a hundred years earlier. Meanwhile, his contemporary Andrew Carnegie moved from Scotland to Pittsburgh and built ironworks of a different kind. Big steel supplanted craft workshops. By the time John William retired at the beginning of the twentieth century, the old craft industries in the north of England were dying. In the next generation, young boys who wanted to get ahead did not become blacksmiths. Many of them, like my father, moved south and went to college.

And still, the human heritage that gave us toolmaking hands and inquisitive brains did not die. In every human culture, the hand and brain work together to create the style that makes a civilization. In every civilization, the skilled artificer has an honored place beside the scribe and the shaman. Our own civilization is no exception.

During the first half of the twentieth century, as the young people of the next generation forgot the skills of my grandfather, they learned new skills and started new industries. They built radio transmitters and receivers, microscopes and telescopes, motorbikes and flying machines. They bred hybrid corn and new varieties of flowers and fruit. Each of these industries grew out of small beginnings and flourished as a craft industry before evolving into large-scale organization and mass production. The early years of the century were the golden age of radio and of flying machines, when inventors could build with their own hands machines that would change the world.

As we moved into the second half of the twentieth century, it seemed that craft industries were dwindling. Mass production dominated the new technologies of television and synthetic materials and large-scale agriculture. Young people, it seemed, had only two choices: either to join the ranks of employees of big enterprises, or to lose interest in technology altogether. The third alternative, to make a living as an artificer with a skilled craft, was no longer practical. But then science emerged to fill the gap.

I remember vividly a scene from the 1960s, when young people were at their most rebellious and technology most unpopular.

Bare feet and outrageous behavior were the prevailing fashions among students. I happened to walk into a basement workshop in the physics building at Cornell University. There I saw two students, dressed in the customary style, with bare feet and long unkempt hair. They were working with intense concentration, building a cryostat, a superrefrigerator for low-temperature experiments using liquid helium. This was not an ordinary helium cryostat that would take you down to one degree above absolute zero. This was a new type of cryostat, working with the rare isotope of helium, that would take you down to a few millidegrees above absolute zero. The students were exploring a new world and a new technology. The working volume of the cryostat was extremely small. It had to be surrounded with many layers of vacuum-tight insulation, and it had to be connected to the outside world by a network of tiny tubes and wires. The students were absorbed in putting this intricate maze of tubes and wires together.

Their brains and hands were stretched to the limit. They had to be sure that every joint was tight, every wire in exactly the right place. I do not remember their names. I do not know whether they stayed on at Cornell and became professional physicists. If they did, it is possible that one or both of them won a Nobel Prize thirty years later, when three Cornell physicists shared the prize for the discovery of superfluidity in the rare helium isotope. At the time when I saw them as students putting the apparatus together, they were not dreaming of Nobel prizes. They were driven by the same passion that drove my grandfather, the joy of a skilled craftsman in a job well done. Science gave them their chance to build things that opened new horizons, just as their ancestors built ships to explore new continents. They had found a creative middle way, between the hierarchical world of big business and the utopian dreams of student rebellion.

During the last fifty years, science has given birth to a new golden age of craft industry. As science extends its reach, it needs new instruments that are increasingly delicate and precise, and it trains students and technicians to build them. Wherever experimental science is done, young men and women are learning to build instruments, using the new materials and new concepts that science has made available. Then the crafts that were nurtured in the laboratory find uses in the world outside. The same young people launch start-up companies to manufacture and sell instruments to other users. And a new craft industry grows. Around every large center of scientific research we find a swarm of craft industries. Silicon Valley grew around

Stanford, the Route 128 corridor around Harvard and MIT, the US1 corridor in New Jersey around Princeton and Rutgers. Some of the new little enterprises outgrow their origins and become large-scale manufacturing enterprises. Other small companies with newer crafts emerge to take their places.

One of the most important tools of modern science is the computer. The building of computers began as a craft industry, John von Neumann supervising an unruly gang of mathematicians and engineers at the Institute for Advanced Study in Princeton, other similar gangs building one-of-a-kind machines elsewhere. Tracy Kidder in his book *The Soul of a New Machine* portrayed the spirit of a craft industry still surviving into the modern era. But the machine that Kidder described and the team of engineers who designed it did not survive. The machine was a failure; it could not compete with the new workstations produced by the Sun Corporation, and the team that built it dissolved. A few years later, Seymour Cray and Danny Hillis, the last of the old breed of craftsmen manufacturing computers as independent entrepreneurs, admitted defeat. They could not compete in the marketplace with the big producers. The craft industry era of computer manufacture is now coming to an end.

But science gave birth to a larger craft industry that is still flourishing, the software industry. Computer software began as a tool of science and later spread to all areas of industry and commerce. Wherever serious computing was done, young people learned to write software and to use it. In spite of the rise of Microsoft and other giant producers, software remains in large part a craft industry.

Because of the enormous variety of specialized applications, there will always be room for individuals to write software based on their unique knowledge. There will always be niche markets to keep small software companies alive. The craft of writing software will not become obsolete. And the craft of using software creatively is flourishing even more than the craft of writing it. The availability of software and the skill to use it gave rise to a whole constellation of new craft industries, from desktop publishing to computer-aided design and manufacturing. A large fraction of the small businesses now operating owe their existence to software that in turn owes its existence to science. This is true even of the more frivolous applications such as computer-aided dating services and computer-aided horoscopes.

Another constellation of craft industries grew out of biology. My youngest daughter, needing to improve her financial situation before entering medical school, worked for a small company preparing DNA libraries. The libraries are collections of standardized samples of DNA. They are sold to laboratories engaged in genetic research or medical diagnosis. For example, if somebody has a bacterial infection or a fungus infection that is hard to diagnose, you can use a library of bacterial DNA or fungus DNA prepared from known species to identify the unknown species that is causing the infection. A small biotechnology company cultures the bacteria and funguses and extracts the DNA to make the libraries. This is a typical craft industry of the modern age. The libraries made by a specialized company are cheaper and more reliable than libraries made by the users for themselves. The employees of the company take pride in their work, quality control being the most important part of it. They ship their libraries all over the world, like the Yorkshire boilermakers a hundred years earlier. My daughter enjoyed the work and showed a talent for it, helped by a few of her great-grandfather's chromosomes.

Science is constantly changing, and the craft industries that it engenders must change too. Technologies rise and fall, and fashions come and go. In the future, many of the small enterprises of today will be consolidated, and new small enterprises will have to find different niches to fill. Today the most successful craft industries are concerned with software and biotechnology. The craft industries of the future might be concerned with neurophysiology or ecology, with technologies not yet invented or with sciences not yet named. Two facts of life will not change. Science will continue to generate unpredictable new ideas and opportunities. And human beings will continue to respond to new ideas and opportunities with new skills and inventions. We remain toolmaking animals, and science will continue to exercise the creativity programmed into our genes.

*   *   *

Some scientific revolutions arise from the invention of new tools for observing nature. Others arise from the discovery of new concepts for understanding nature. Two historians, Peter Galison and Thomas Kuhn, have explored in depth the process of scientific discovery in the modern age. Galison's great work, *Image and Logic*, was published in 1997. Kuhn's *The Structure of Scientific Revolutions* appeared thirty-five years earlier. Kuhn died before he had a chance to say what he thought of *Image and Logic*. Galison and Kuhn were both trained as physicists before they became historians. Both of them are primarily interested in the history of physics, and both have mastered the technical details of physics as well as the scholarly craft of historiography. Yet their views of the history of science are totally different. Their two books have almost nothing in common. Galison's book contains hundreds of pictures of scientific apparatus; Kuhn's book contains only words. For Galison the process of scientific discovery is driven by new tools, for Kuhn by new concepts. Both pictures are true and neither is complete. The progress of science requires both new concepts and new tools.

The difference between Galison and Kuhn is largely a difference of emphasis. Kuhn emphasized ideas and Galison emphasizes things.

Unfortunately, Kuhn's version of history was dominant for thirty years before Galison's version appeared to restore the balance. Kuhn's book became a classic and gave to his nonscientist readers a one-sided view of science. Kuhn wrote about the battles between rival concepts, and some of his readers were left with the impression that science is a largely subjective affair, a struggle between conflicting human viewpoints, rather than an objective struggle between the precision of tools and the ambiguities of nature. Kuhn saw science from the point of view of a theoretical physicist, taking the experimental data for granted and describing the great leaps of theoretical imagination that enable us to understand it. Galison sees science from the point of view of an experimental physicist, describing the great leaps of practical ingenuity and organization that enable us to acquire new data. Although I am a theorist, I happen to find Galison's view of science more congenial. Most theoretical physicists have an opposite bias, having more respect for philosophy and less respect for gadgetry. The science that I describe in this book is Galisonian science, based on the clever use of tools rather than on philosophical argument. Science for me is the practice of a skilled craft, closer to boiler-making than to philosophy.

Most of the recent scientific revolutions have been tool-driven, like the double-helix revolution in biology and the big-bang revolution in astronomy. But concept-driven revolutions still sometimes occur. A good example of a recent concept-driven revolution is the plate tectonics revolution of the 1960s, when the international community of geologists rather suddenly became converted from the belief that continents are immovable to the belief that continents float around on moving plates of rock. The evidence for moving continents had been carefully assembled long before by the German meteorologist Alfred Wegener. Wegener presented the evidence in a book published in 1915, but the evidence was disregarded for fifty years because it contradicted the prevailing dogma. There was then no consistent conceptual framework into which moving continents could fit. The prevailing dogma quickly collapsed in the 1960s when a new conceptual framework was invented, with continents attached to moving plates composed of oceanic crustal rock. The new concept had each plate emerging out of mid-ocean ridges at one side and sinking into ocean trenches at the other. This revolution followed exactly the pattern described by Kuhn in his book. Facts that seemed irrelevant when they disagreed with the accepted concepts suddenly became important when the concepts changed so that the facts could be understood. The new concept of plate tectonics made everything look different. After the revolution, it was difficult to understand how Wegener's evidence could have been so long ignored. According to Kuhn, a concept-driven revolution happens when a whole generation of scientists suddenly responds to long-neglected evidence contradicting accepted dogmas. The response is sudden because it has to wait until a new concept emerges.

It often happens when a new religion is founded that the followers become far more rigid and doctrinaire than the founder. This happened among the followers of Thomas Kuhn. Kuhn himself never intended to found a new religion. But a group of philosophers and historians of science call themselves Kuhnians and propound views far more extreme than his. Some Kuhnians have asserted that the acceptance of new scientific theories is based on political and economic power struggles rather than on scientific evidence. A few years ago, I happened to meet Kuhn at a scientific meeting and complained to him about the nonsense that had been attached to his name. He reacted angrily. In a voice loud enough to be heard by everybody in the hall, he shouted, "One thing you have to understand. I am not a Kuhnian." Kuhn never said that science is a political power struggle. If some of his followers claim that he denied the objective validity of science, it is only because he overemphasized the role of ideas and underemphasized the role of experimental facts in science. He started his career as a theoretical physicist. If he had started as a biologist, he would not have made that mistake. Biologists are forced by the nature of their discipline to deal more with facts than with theories.

The debate about the objective validity of science was recently enlivened by a successful practical joke. One of the professional journals in the social analysis of science has the name *Social Text*. It publishes papers written by the more extreme believers in the doctrine that science can be understood as a social construction without paying much attention to its subject matter. A theoretical physicist at New York University, Alan Sokal by name, was teaching a course explaining the elements of science to students of the humanities. He discovered to his dismay that many of the students in his class, and the professors who were teaching them humanistic subjects, actually believed the stuff they were reading in *Social Text*. So he decided to use the weapon of satire to demolish the journal's pretensions. He wrote a satirical paper with the title "Transgressing the Boundaries: Towards a Transformative Hermeneutics of Quantum Gravity," imitating perfectly the style and the political agenda of *Social Text*. The political agenda is beautifully captured in Sokal's final paragraph. Here are a few sentences from his peroration:

Mainstream Western physical science has, since Galileo, been formulated in the language of mathematics. But whose mathematics? The question is a fundamental one, for, as Aronowitz has observed, neither logic nor mathematics escapes the contamination of the social. And as feminist thinkers have repeatedly pointed out, in the present culture this contamination is overwhelmingly capitalist, patriarchal and militaristic. Thus, a liberatory science cannot be complete without a profound revision of the canon of mathematics.

The paper sounded like a typical *Social Text* article but was actually a farrago of nonsense. Even the title makes no sense. Quantum gravity is an esoteric branch of theoretical physics, far removed from any possible connection with social or political concerns. Sokal submitted the paper to the editors of *Social Text*, and they published it, demonstrating that they were unable to tell the difference between sense and nonsense. This was a famous victory, the practical joker outwitting the panjandrums.

In the same year that Sokal played his joke, Peter Galison published his great work *Image and Logic*, providing a more solid corrective to the portrayal of science as a political power struggle. Galison portrays science from the point of view of the toolmaker. At the beginning of his book, Galison says:

This is a book about the machines of physics.... Pictures and pulses—I want to know where they came from, how pictures and counts got to be the bottom-line data of physics. The Göttingen number theorist Edmund Landau, on hearing of something too applied for his taste, looked over in disgust and pronounced it *Schmieröl*—engine grease. But that is what I want to know about: how all these machines, these gases, chemicals, and

electronics, come to make facts about the most theoretically articulated quadrant of nature. Histori-
cally, historiographically, philosophically, this book
is a back-and-forth walk through physics to explore
the site where engine grease meets up with experimental results and theoretical constructions.

Galison does not underestimate the importance of
political power struggles. Many of the most dramatic
passages in his book describe power struggles, with the
titans of experimental physics fighting for the money
and turf that they need in order to build their tools and
do their experiments. But the power struggles
described by Galison are different from the power
struggles portrayed in *Social Text*. The Kuhnians write
about ideological struggles, right against left, capitalist
against socialist, masculine against feminine. Galison
writes about practical struggles, battles to decide whose
machine gets built, whose detector gets used, whose
theory gets tested. After the human battles are won
and lost, the experiment goes forward and the result is
determined by tools and nature. If the tools are bad,
nature's voice is muffled. If the tools are good, nature
will give a clear answer to a clear question. The result
of the experiment depends on nature and on the quality of the tools, not on the ideology of the experi-
menter.

One of the heroes of Galison's story is Marietta
Blau, a physicist who worked at the Radium Institute
in Vienna from 1923 to 1938. She was born into a cul-
tured Jewish family but did not enter the family business of music publishing. Instead she became the pioneer of the art of using photographic emulsions as a
tool for detecting elementary particles. She began with
the standard commercial X-ray film used by dentists
and gradually improved the quality and sensitivity of
the emulsion until it could record a visible track when

a single high-energy particle passed through it. In her hands, the emulsion became a powerful tool for elementary particle physics. Compared with the cloud chamber, the previously dominant tool, the emulsion had two advantages. Particle tracks in emulsion were more precisely localized and less diffuse than tracks in a cloud chamber. And the emulsion remained sensitive over long periods of time, whereas the cloud chamber was sensitive only for a fraction of a second. For a few years after World War II, the photographic emulsion became a dominant tool, used by Powell in England to discover new particles that revolutionized the field of particle physics. Meanwhile, Marietta Blau's work was disrupted when Hitler annnexed Austria and she was driven into exile in Mexico. She never had an opportunity to reap the harvest of discoveries that her emulsions made possible. At the end of her life she returned to Austria, where a surviving brother could take care of her. The photographic emulsion was soon superseded by the electronic detector as the dominant tool of particle physics, but it remains a useful tool for special purposes. Hitler had destroyed Blau's professional career, but he could not destroy the tools that she created or the scientific revolution that her tools brought about.

The Kuhnians have applied their social analysis to biology as well as to physics, but biology is less vulnerable since biologists are closer to the nature that they are studying. Sometimes the social analysis of biology and medicine can be illuminating. To give one example, Emily Martin has looked at the human immune system as reflected in the popular culture of the last hundred years and has found striking connections between immunology and the way ordinary people talk and think. Some less perceptive social analysts have attempted to show that a virus is a social construction. It is difficult to make a strong case for the unreality of a virus. If some virologist had submitted a paper to the editors of *Social Text* with the title "Transgressing the Membranes: Towards a Transformative Hermeneutics of the Viral Genome," the editors would probably have seen the joke and rejected the paper. Virologists are working close to nature. Medical scientists generally stay close to nature and do not pay much attention to philosophers.

The virus is a tool, not a theory. And it is a tool for the practice of medicine as well as for the advancement of science. Sixty years of brilliant work by experimental virologists have taught us a great deal about the structure of viruses. We know pretty well what viruses are made of. We know much less about their function. Since their function involves invading and subverting and destroying bigger creatures, we cannot understand the function of viruses without also understanding the function of the bigger creatures on which they prey. This makes the virus an excellent tool for exploring the bigger creatures. To infect a cell with a virus is nature's way of doing invasive surgical intervention. If nature had not already invented viruses, we would have been forced to invent them ourselves. As tools, they have all the virtues that are rarely found together, standardization combined with variety, reproducibility combined with malleability, ruggedness combined with specificity, precision combined with cheapness.

One of the scientific revolutions of the future is likely to arise from the technology of artificial viruses. It is already possible to synthesize an entire virus from its chemical components, given a knowledge of the sequence of bases in its genes. As we acquire a detailed understanding of the function of viruses and the cells that they invade, the way will be open to use synthetic viruses as tools for the exploration and manipulation of cells. Synthetic viruses could be designed to enter a cell and turn particular metabolic processes on or off so that we can observe the consequences.

Synthetic viruses might help to elucidate in detail the pathways by which cells become malignant. And one may dream of a synthetic virus designed to be sensitive to the difference between healthy and malignant cells, so that it is harmless to healthy cells and gives to malignant cells a signal to destroy themselves. The synthetic virus gives us realistic hope of replacing the surgeon's knife and the chemotherapeutic poison by a more discriminating tool.

*   *   *

The genetic engineering of viruses and other creatures will give rise to new scientific revolutions that will have profound effects on human life. Meanwhile new tools of observation will be causing scientific revolutions in astronomy. I find it illuminating to look at biology and astronomy together, in the hope that each of these disciplines may have something to learn from the other. I will tell stories about the sky and the genome, describing some scientific revolutions that already happened and others that we can already see on the horizon. The sky pulls us outward to explore the universe, the genome pulls us inward to explore the workings of our minds and bodies. Here are two stories of strange planets and strange viruses, stories of two pioneers exploring the sky and the genome and finding weird objects that nobody expected.

The plurality of worlds was a central question for theological inquiry, long before it became accessible to scientific observation. In the sixteenth century, Giordano Bruno argued that the greatness of God was better demonstrated in the creation of many worlds than in the creation of only one. He preached this doctrine all over Europe, to Catholics and Protestants, and found in many places a favorable response. Only when he made the mistake of returning to his native Italy did he run into serious trouble. He was dragged to Rome, condemned by the Inquisition as a heretic, and burned at the stake.

But his death did not put an end to his doctrines. Gradually, as the Catholic and Protestant churches became more tolerant, the plurality of worlds ceased to be a mortal heresy and became an acceptable subject for sermons and hymns. In the eighteenth century, when the immense scale of the astronomical universe was revealed by William Herschel's telescopes, liberal churchmen were not slow to claim the millions of worlds in the sky as evidence of God's greatness and to imagine the millions of stars to be accompanied by millions of planets. In the twentieth century, belief in the plurality of worlds has become orthodox. Since our sun appears to be a typical star, it seems reasonable to expect that other stars will be accompanied by families of planets.

For sixty years, beginning in the 1930s, astronomers searched in vain for evidence of planets around alien stars. Then suddenly, in 1992, the first family of extrasolar planets was discovered by Alexander Wolszczan, a young radio-astronomer who teaches at Pennsylvania State University. The news of his discovery was greeted by the community of astronomers in Princeton with considerable skepticism. I was lucky to be present when Wolszczan came to Princeton to confront the skeptics. This was a historic occasion. It provides an excellent example of the way the scientific establishment deals with young people who claim to have made important discoveries.

Alexander Wolszczan sat down to lunch with about fifty astronomers gathered together from Princeton and neighboring communities. The proceedings were informal and superficially friendly, but there was high tension in the air. Wolszczan had claimed to find planets orbiting around the wrong kind of star, not a normal star like the sun but a tiny, rapidly spinning object known to astronomers as a millisecond pulsar. A millisecond pulsar is as different from our sun as it is possible to be. It is a collapsed remnant, only a few miles in diameter, formed when a giant star explodes at the end of its life.

The explosion drives off into space the massive envelope of the star, leaving behind only the tiny collapsed core. If the giant star had possessed a family of planets before the explosion, it was unlikely that any of them would have survived and remained attached to the remnant. It was equally difficult to imagine new planets being formed out of the debris from the explosion, since the debris is expelled from the star with enormous velocity. The prevailing view among astronomers was that a millisecond pulsar was the least likely type of star to be accompanied by planets. Wolszczan himself shared this view. The last thing he expected to find when he observed his pulsar was planets. He was not searching for planets. He came to believe that he had found planets only after all other explanations of his observations had failed.

The evidence for Wolszczan's planets was highly indirect. The only thing that he measured directly was the timing of the pulses of radio-waves emitted by the pulsar as it rotated several hundred times per second. He measured the timing of the pulses with extraordinary accuracy, made possible by new tools, new clocks, and new computer programs. The pulsar itself is a clock, its pulses clicking with a regularity comparable to the most accurate man-made clocks. But Wolszczan, comparing the pulsar clock with his atomic clocks to microsecond accuracy over several years, found strange discrepancies. The timing of the pulses wandered, sometimes speeding up and sometimes slowing down, in a way that could be explained as the disturbing effect of two planets orbiting around the pulsar. From the detailed pattern of the wandering,

Wolszczan could deduce the orbits of the planets and could calculate that they each must weigh about three times as much as the Earth.

The most important of the new tools that led Wolszczan to his discovery were software programs. The cutting edge of astronomical technology today is more concerned with software than with hardware. The telescope that Wolszczan used for his observations, the great radio-telescope at Arecibo in Puerto Rico, was forty years old. The software that he used to extract pulsar signals from the telescope was new. The pulsar that showed the irregular timing is extremely faint, and the irregularity makes it even harder to detect than a normal millisecond pulsar. Only by using specially designed software programs was it possible to disentangle the faint modulation of the pulsar from the roar of radio noise in which it is buried. The credibility of Wolszczan's discovery depended entirely on the credibility of his software programs. In order to convince his colleagues that his planets were real, he had to convince them that his software programs were free from bugs.

The first thing that a good scientist does when confronted with an important discovery is to try to prove it wrong. Before telling anybody else about it, Wolszczan tried very hard to prove himself wrong. He checked and rechecked his apparatus and his computer programs, he searched for possible sources of interfering radio signals, he examined every instrumental effect that might have misled him. Only after all attempts to prove himself wrong had failed did he announce his discovery. As a result, when he came to Princeton to have lunch with the assembled astronomers, he was well prepared. Each astronomer who doubted the reality of Wolszczan's planets took a turn as prosecuting attorney, asking sharp questions about the details of the observations and trying to find weak points in Wolszczan's analysis. Wolszczan answered each question calmly and completely, showing that he had asked himself the same question and answered it long before. At the end of the lunch there were no more questions.

Wolszczan came through the ordeal victorious, and the skeptics gave him a friendly ovation. That is the way the scientific process is supposed to work, Wolszczan's planets proclaiming their existence against all theoretical expectations, Wolszczan's skill and integrity overcoming all opposition from his more senior colleagues.

After the discovery of the two larger planets had been published and generally accepted, Wolszczan continued to study the pulsar signals and announced the discovery of a third, much smaller planet. The small planet appeared to be orbiting close to the pulsar with a short orbital period. The period was observed to be 25.34 days. Unfortunately, this is precisely equal to the rotational period of the sun. As the sun rotates, it carries around with it a pattern of extended magnetic fields and solar winds that can influence the timing of pulsar signals. A periodic modulation with the same period was recently found in the radio signals from the Pioneer 10 spacecraft, which is traveling away from the sun far beyond Pluto. The modulation of the signal from Pioneer 10 is certainly due to the sun's rotation. This makes it likely that the modulation that Wolszczan observed in the pulsar signal with this period was also due to the sun. So the third planet probably does not exist. This mistake shows how easy it is for even a first-rate and careful observer to deceive himself. But the two larger planets are real, and the disappearance of the small planet does not make the discovery of the two larger ones less important.

During the five years since Wolszczan's discovery, ten planets have been found by other astronomers. In astronomy as in other areas of science, it is always easier to discover additional examples of a new type of object after the first one is found. The astronomers who made the later discoveries had less skepticism to overcome than Wolszczan. The later discoveries differ from Wolszczan's in two other ways. First, the planets belong to ordinary stars like the sun and not to millisecond pulsars. Second, the planets have much larger masses, several hundred times the Earth's mass, unlike Wolszczan's planets, which have only three times the Earth's mass. The newer planets are giants like Jupiter rather than dwarfs like our Earth. It was inevitable that the newer planets would have big masses. For a planet around a normal star to be detectable, it must have a mass comparable with Jupiter's. The most interesting planets from a human perspective, those that have a mass like Earth's and belong to a star like the sun, are undetectable with existing instruments. Planets with earthlike mass can at present be detected only if they belong to millisecond pulsars. It is no accident that the only planets found until now with earthlike mass are those discovered by Wolszczan. Almost all astronomers believe, after the recent discoveries, that planets with earthlike mass orbiting stars like the sun are abundant in the universe. But such planets will not be detected until new instruments or new techniques of observation are developed.

Wolszczan's visit to Princeton reminded me of an equally dramatic occasion fifteen years earlier, when the visitor was Fred Sanger.

Fred Sanger is at home in the genome as Alexander Wolszczan is at home in the sky. Sanger is a molecular biologist who created the technical tools to find the sequence of bases in long stretches of DNA. He is one of the great tool-builders of our time. I have a vivid memory of Sanger coming to Princeton and talking about his sequencing of the 5375 bases in the genome of the φ X174 virus. That was about twenty years ago, in 1977 or 1978, when the sequencing of five thousand bases was ſtill a heroic enterprise. He described how he had done the sequencing and how the genes were arranged on the DNA. The amazing thing about this genome was that the genes overlapped. Let me explain what it means for genes to overlap. A gene is a long ſtring of bases that work like a written message, telling us how to make proteins, which are the essential conſtituents of all living cells.

The ſtring of bases can be thought of as a ſtring of letters of the alphabet, for example, ABCDE-FGHIJKLMN. But the DNA in this virus is a closed ring, so you muſt imagine the ſtring of bases to be a closed cycle of 5375 letters without any beginning or end. You muſt imagine that there is a long ſtring of letters before the A and another long ſtring of letters after the N. A typical gene contains many hundreds of letters. The cell reads the letters three at a time to translate them into a protein molecule. This means that the cell muſt choose where to begin reading. If the reading begins at A, the protein will be specified by the triplets ABC,DEF,GHI,JKL and so on. If the reading begins at B, the protein will be specified by the triplets BCD,EFG,HIJ,KLM and so on. If the reading begins at C, the protein will be specified by the triplets CDE,FGH,IJK,LMN and so on. The three ways of reading are called reading-frames. For a given ſtring of letters, the three reading-frames will give totally different proteins. The DNA also contains special sequences of letters called initiation sequences, one for each gene, which ſtand in front of the gene and tell the cell where to begin reading.

Before Fred Sanger sequenced his virus, every biologiſt took it for granted that each ſtretch of letters on the DNA could be read in only one way. For example, if the triplets ABC,DEF,GHI,JKL could be translated into a protein that made sense, then it seemed wildly improbable that the triplets BCD,EFG-HIJ,KLM or CDE,FGH,IJK,LMN could also make sense and be translated into useful proteins. Everyone took it for granted that the DNA would contain an initiation sequence that would fix the reading-frame uniquely. If the initiation-sequence said, "Begin at A," then that was it. A cell that made a mistake and began reading at B would only produce garbage. But Sanger told us that in his virus he found long stretches of DNA that were coding for one protein in one reading-frame and simultaneously coding for a different protein in another reading-frame. There was even a short stretch that was coding in two reading-frames and at the same time functioning as an initiation sequence for a third gene. How could that possibly work? How could a sequence make sense in different reading-frames simultaneously? I found it difficult to believe the virus could be as clever as that. But then I remembered Mozart. It happened that at that time my youngest daughter was ten years old, and she was studying the violin. One of the pieces that she was studying was a duet for two violins. I also played the violin, not as well as she did, but well enough to play a simple duet with her. The remarkable thing about this particular duet is that it is written with both parts on a single page of music. The players have to stand on opposite sides of a table with the music lying flat between them. One player plays the music right-side up from the top down, while the other plays it upside-down from the bottom up. And it sounds good. The two violins blend well, and there is some interesting counterpoint. It is not a great piece of music, but it makes sense. Even the sharps make sense, although the sharp signs are operating on different notes when you turn the page upside down. Mozart could not use flats because flat signs do not have rotational symmetry, so he wrote it in the key of G major, which does not require flats. He must have enjoyed himself when he composed it. I doubt whether anybody other than Mozart could have done it. I like to call Fred Sanger's virus the Mozart virus. It shows that nature can compose a genome as cleverly as Mozart could compose a duet. Like the duet, the Mozart virus is not a great work of art, but it is a great example of nature's ingenuity.

Molecular biology and astronomy are the most rapidly growing areas of science today, just as physics was the most rapidly growing science fifty years ago. The stories of the Wolszczan planets and the Mozart virus illustrate the way surprises in science often arise from new tools rather than from new concepts. Sanger did not have any new concept in mind when he decided to study the φ X174 virus. It was a well-known virus, infecting the bacteria called E. coli that live peacefully in the human intestine and help to digest our food. The virus had been seen in pictures taken with an electron microscope. It had the shape of an elegant little regular icosahedron with twenty triangular faces. Its genes had already been identified and roughly mapped by genetic methods. Sanger studied this well-known virus with a new tool, the plus-and-minus method that he invented for sequencing DNA.

He chose to sequence the virus because it was a good exercise for his new tool, a complete genome that was small enough to be within his reach. He was using the new tool to explore new territory, without having any preconceived theory of what he would find there. The fact that the virus had overlapping genes was as much of a surprise to Sanger as it was to everybody else. Likewise, the fact that a millisecond pulsar could have planets was a total surprise to Wolszczan. These are typical examples of discovery in science. Most discoveries are stumbled upon, not planned. The discoverer usually does not know until afterwards whether the discovery is important or unimportant. The

Mozart virus turned out to be unimportant and the Wolszczan planets turned out to be important, but the ways in which Sanger and Wolszczan made the discoveries were not very different. Important discoveries are only recognized as important after they have given rise to scientific revolutions.

* * *

There are now two big science projects in progress of central importance to biology and astronomy: the human genome project and the Sloan Digital Sky Survey. They are applying the tools of industrial-scale organization and data processing to the genome and to the sky. The tools of astronomy have tended to be large and expensive, ever since William Herschel built his mammoth reflecting telescope to explore the universe beyond the solar system two hundred years ago. Wolszczan discovered his planets with the radio-telescope at Arecibo, the largest single-dish telescope in the world. The tools of biology are smaller, but they also show a tendency to become large and expensive as biologists penetrate more deeply into the architecture of living cells. Fred Sanger was working in a laboratory with modest equipment when he sequenced the ϕ X174 virus in 1977. He is now Sir Fred Sanger, director of a large enterprise dedicated to sequencing genomes on a massive scale. His enterprise is an important part of the international effort to sequence the human genome.

The goal of the human genome project is to read and record the entire string of three billion letters in the DNA of a human being, in the same way as Fred Sanger read the string of five thousand letters in the DNA of his virus twenty years earlier. The project is very slow and very expensive. It will take roughly fifteen years and three billion dollars to sequence a single genome. The technology of sequencing has been improved in detail but not radically changed since 1977. The main innovation since 1977 has been the polymerase chain reaction, a clever chemical trick that allows us to multiply a single molecule of DNA so that we can make a huge number of identical copies of it for sequencing. The basic process of sequencing is still wet chemistry. This means that the DNA to be sequenced is exposed to various chemical reagents dissolved in water, the reaction products are allowed to flow onto a tray covered with gelatine, and high voltage electric fields pull the reaction products along the tray. The pattern of the reaction products determines the sequence. These operations are now done by automatic machines instead of by human hands, but the process is still time-consuming, labor-intensive, and frequently unreliable.

The human genome project began in 1990 and is supposed to be finished in 2005. In 1999 more than half the time has passed but less than a tenth of the genome has been sequenced. The human genome, unlike the viral genome, uses only a small fraction of its DNA for coding proteins. The virus has no space to spare and must pack its DNA tightly with genes, even forcing some of the genes to overlap. The human cell has plenty of space and can afford to be wasteful. Only about 3 percent of the human genome codes are for proteins. Even if we include a generous allowance for control sequences acting to switch the genes on or off, the coding and control sequences together are less than 10 percent of the genome. The remaining 90 percent of the DNA in the genome has no known function, and nobody knows why this 90 percent is there. Some biologists call it "Junk DNA" and suppose it to be parasitic, maintained by the processes of DNA replication without providing any benefit to the organism. More likely, it has some useful function, but the function may be purely structural. It is possible that the detailed sequence of 90 percent of the genome has no biological significance. This being the case, the great majority of biologists engaged in sequencing human DNA have concentrated their attention on the coding regions, whose function is known to be crucial. They have sequenced the entire genomes of simpler creatures, viruses and bacteria and protozoa, whose genomes are filled with genes and have little space for noncoding regions. They have sequenced many of the coding regions of mice and fruit flies in addition to humans. The object of their game is to identify and analyze genes. It is the genes that are medically important. Genes make us sick or healthy, sane or crazy, fat or thin. So far as the medical applications of sequencing are concerned, there is no reason to be in a hurry to sequence the 90 percent of the human genome that has no obvious connection with genes.

In my opinion, the decision of the administrators of the genome project, to finish the sequencing of the entire human genome by the year 2005 using the existing methods, was unwise. The decision was driven by politics and not by the needs of science and medicine. Since the project had been launched with the declared objective of finishing the sequence within fifteen years, finishing it on time became politically imperative. Because the existing methods of sequencing are slow and expensive, the decision to finish it by 2005 requires that the lion's share of available resources will be devoted to the mass production of sequence. Research directed towards new sequencing methods that might be faster and cheaper will be curtailed. The project will no longer support the development of sequencing tools that will not be ready for large-scale use before 2005. It would have been better for science if the administrators had decided to continue using the existing technology to finish sequencing the scientifically important part of the genome and meanwhile to increase the effort to find better methods for sequencing the rest of it. In this way, the medically significant part of the project would have been completed by 2005 at lower cost, and the money saved could have been invested in tools for much cheaper sequencing in the years after 2005. In science, to change the objectives of a program in the light of new discoveries is a sign of wisdom. In politics, it is a sign of weakness. Unfortunately, politics prevailed over science.

The Sloan Digital Sky Survey is a quicker and cheaper project, organized as a cooperative enterprise by seven universities, including Princeton. Its purpose is to make a precise electronic map of the northern half of the sky. It will observe the entire northern sky with electronic cameras and record the images in digital memory. This will take roughly five years and cost roughly eighty million dollars, one-third of the duration and one-fortieth of the cost of the human genome project. If all goes well, the southern sky will be done later in the same way. The northern sky will be observed by a single telescope at Apache Point in New Mexico, built for the purpose and dedicated to the project for five years. The telescope began operations in 1998. The Sloan Foundation paid with private money for the telescope and for part of the cost of instruments and operations. The government is paying for the rest of it.

The most difficult part of the project is not the optical and electronic hardware but the data-processing software. The survey is done by recording each patch of sky with electronic cameras in five colors, so that every patch in the sky is described by a string of numbers in the digital memory. For each patch there are five accurate measurements, recording the intensity of light coming from that direction in each of the five colors. In addition, the survey will record spectra and measure red-shifts of a million galaxies and other distant objects. The data in the memory will be made available to every astronomer who has the equipment to use it. The output of the survey, like the output of the genome project, will have three main uses. First, an astronomer searching for rare kinds of object to study, like a biologist searching for rare kinds of genes, can find them quickly by searching the database with electronic search algorithms. Second, an astronomer who is interested in some particular region of the sky, like a biologist who is interested in some particular gene, can obtain information about the local environment. Third, the survey will provide statistical information about the distribution and behavior of every kind of object in the sky, just as the genome project will provide statistical information about the distribution and behavior of every kind of gene. Beyond these three predictable uses, each project will provide a database that will be used in unpredictable ways to organize our knowledge of the universe and of ourselves. Astronomers will browse in the sky-survey database to find their way around the sky, and biologists will browse in the genome database to find their way around the chromosomes. The design of efficient browsing software will be an important part of both projects.

The sky survey and the genome project are both scientifically splendid. Each of them is a major step forward in science. But there is one glaring difference between them: the factor of forty in cost. The difference in cost is crucial, because it implies a difference in sustainability. A project is sustainable if it is cheap enough to be the first of a series continuing indefinitely into the future. A project is unsustainable if it is so expensive that it cannot be repeated without major political battles. A sustainable project marks the beginning of a new era. An unsustainable project marks the end of an old era. The most famous example of an unsustainable project was the Apollo program of manned expeditions to the moon. After it was decided that the Apollo program was too expensive to continue, the Apollo hardware was scrapped and the entire space program was forced to make a fresh start. Because it was unsustainable, the Apollo program set back the development of space science by twenty years.

Our digital sky survey is clearly sustainable. After it is finished, it will probably be repeated many times with variations and improvements, reaching out to fainter objects and measuring colors with higher resolution. It will set the style for future sky surveys of many kinds. The human genome project at the current cost is not sustainable. After it is finished, it is unlikely to be repeated until cheaper methods of sequencing become available. The enormous potential rewards of the genome project will not be reaped until large-scale sequencing becomes cheap enough to be sustainable.

To advance the science of medicine, it is not enough to sequence a single genome. We shall need to sequence a multitude of genomes belonging to people of many types with different medical histories. To advance the science of biology we need to sequence not only humans but thousands of other species. We need ultimately to sequence the genome of the entire biosphere. The biosphere contains millions of species, and an average species has a genome containing millions of bases. A biosphere genome project would require the sequencing of trillions of bases. The total amount of data storage required to record the genome of the entire biosphere is surprisingly small, only a little larger than the database of the Sloan Digital Sky Survey. The genome of the biosphere could be recorded on compact disks that would fit into a standard filing cabinet. But the process of extracting the data from DNA is almost a million times slower than the process of extracting data from the sky. Before a biosphere genome project can be considered as a practical possibility, the cost of sequencing must be reduced, and the speed increased, by a factor of a hundred or a thousand.

Why is there this huge difference in cost between the sky survey and the genome project? There are many reasons for the difference. One of the main reasons may be the fact that astronomers and biologists have different attitudes towards their tools. Astronomers have traditionally invented and built their own tools. My colleague Jim Gunn, one of the leaders of the sky survey, himself invented and glued together the electronic camera that is the essential observational tool. The first little wide-field telescope that made rapid surveys possible was installed on Palomar Mountain in 1937 by the astronomer Fritz Zwicky. Astronomers usually wait, before beginning a project, until they have the necessary tools in hand. Biologists, with a few notable exceptions such as Fred Sanger, traditionally buy their tools and do not build them. They begin projects using whatever tools happen to be available. Building their own tools is not a part of their culture.

*   *   *

Astronomers all over the world are now developing a new tool called gravitational tomography. The idea is similar to the idea of X-ray tomography, which is the basis of the CAT scan used routinely to look inside patients' bodies. Gravitational tomography looks at the universe in the same way as a CAT scan looks at a diseased lung. The universe is similar to a human body in one respect. Most of the universe, like most of the human body, cannot be seen with ordinary light. Fritz Zwicky discovered more than fifty years ago that clusters of galaxies are held together by invisible mass that is far larger than the mass of the visible galaxies. Zwicky was observing clusters of galaxies systematically, using his little telescope on Palomar Mountain. He observed that the galaxies in clusters move around randomly with high velocities and that if the galaxies were bound only by the gravitational attraction of the visible mass the cluster would quickly fly apart. Because the clusters evidently do not fly apart, the invisible mass must exist. Since the days of Zwicky, many powerful new tools have been introduced into astronomy for the observation of visible objects. We observe visible objects with radio-waves and X-rays as well as with light. Our knowledge of the visible universe has become enormously more detailed and more complete. The observations of Zwicky have been confirmed, and invisible mass is now known to exist in other places besides clusters of galaxies. In the universe as a whole, the invisible mass probably outweighs the visible mass by more than a factor of ten. But the invisible mass remains invisible. The only evidence for invisible mass is the gravitational effect that it produces on visible objects. The idea of gravitational tomography is to use the bending of light-rays in a gravitational field to localize invisible mass, in the same way as a CAT scan uses X-rays to localize a tumor in a lung.

Gravitational tomography detects invisible mass by observing the gravitational focusing of light from visible objects in the same direction at a greater distance. Any mass, whether visible or invisible, forms a gravitational lens and focuses light when it stands between the earth and a visible object. There are two kinds of lenses: macrolenses, which are as massive as galaxies or clusters of galaxies, and microlenses, which are as massive as single stars or planets. There are two possible kinds of invisible mass: discrete massive objects such as cold dead stars and black holes, and diffuse clouds of exotic invisible particles spread over large volumes of space.

Gravitational tomography is the magic tool that

makes the invisible detectable. Invisible mass in the form of particle-clouds can be detected by macrolensing. Invisible discrete masses can be detected by microlensing. Tony Tyson, working at the Bell Laboratories that used to belong to AT&T and now belong to Lucent Technologies, has made a good beginning with the mapping of large extended distributions of invisible matter by means of macrolensing. He observes the gravitational distortion of images of remote galaxies in the background and deduces the distribution and quantity of invisible mass in the foreground. The invisible mass that he detects in this way is usually, but not always, associated with clusters of visible galaxies.

Meanwhile, the astronomer Bogdan Paczyński at Princeton suggested ten years ago that microlensing could be used as a practical tool of exploration, to discover discrete invisible masses in our own neighborhood, using stars in the Large Magellanic Cloud and in the central region of our galaxy as background objects. The Large Magellanic Cloud, discovered by Magellan when he sailed far south to make the first voyage around the tip of South America, is a neighboring galaxy that happens to lie far south in our sky. The Large Magellanic Cloud and the central region of our own galaxy are the two best places in the sky for observing millions of background stars in the field of view of a single telescope. The way to find microlenses is to observe the same field of view once every few weeks to see whether one of the millions of stars has grown suddenly brighter. When a star is seen to grow brighter and then fade back to its original brightness, this may be the result of gravitational lensing by a foreground object passing precisely across the optical path between the star and the telescope. The essential tool that made this method of searching feasible was digital image-processing, which allows millions of star images to be recorded and analyzed in a few minutes. In the laſt three years, four groups of aſtronomers have used the technique with speſtacular success. More than a hundred microlensing events have been seen, and the rate of new discoveries is increasing rapidly. Some of the lensing objeſts are ordinary visible ſtars, but there is evidence that a subſtantial fraſtion of them are invisible. Within a few years, we should have reliable information about the population of discrete invisible objeſts in and around our galaxy.

Another application of gravitational tomography is the search for planets. A planet is a discrete massive objeſt and produces a microlens that may be deteſtable. The duration of a lensing event is roughly proportional to the square root of the mass of the lensing objeſt. Even though the sun weighs three hundred thousand times as much as the Earth, the lensing effeſt of the sun is only five hundred times the lensing effeſt of the Earth. This means that gravitational lensing can be used as a tool for finding planets. Since a ſtar with the same mass as the sun produces a lensing event with a duration of a few weeks, a planet with the mass of the Earth produces a lensing event with a duration of about an hour. Planet lenses would be missed in the exiſting microlensing searches, because the ſtar-fields are not observed often enough to catch them. Penny Sackett, a young American aſtronomer now living in the Netherlands, has found a way to remedy this deficiency. She has set up a projeſt called PLANET, using four small telescopes with good coverage in longitude around the South Pole, one in South Africa, one in Chile, one in Weſtern Auſtralia and one in Tasmania. These are modeſt university telescopes that are not in heavy demand, so they are available with plenty of time for observing and with ſtudents who are happy to collaborate on this projeſt without being paid. Whenever a lensing event is discovered by one of the big microlensing projeſts, the exaſt position of the objeſt is reported to the PLANET observatories, and the students go to work observing the object once every hour. Because the Large Magellanic Cloud is close to the South Pole in the sky, these four observatories around the Pole are able to maintain a continuous watch unless weather conditions are bad for two or three of them simultaneously. Project PLANET began in 1996 and has not yet discovered any planets. If optimistic estimates of the prevalence of planets are true, the project should find a planet once every few years. The advantage of this method of search is that it should work independently of the nature of the massive object to which the planet is attached. It could detect planets around normal stars or around invisible objects of types that are still unknown.

What does the search for invisible mass in the universe have to do with biology? So far as the substance of the science is concerned, there is no connection between gravitational tomography and biology. The connection appears when we consider the style rather than the substance. The style of the work in gravitational tomography is opportunistic, unorganized, spontaneous. Nobody planned it, and nobody administers it from the top. A new tool of observation became available, and a number of bright people grabbed hold of it for a variety of purposes. The money to pay for it comes piecemeal from a variety of sources: the National Science Foundation and the Department of Energy in Washington, the French and Polish governments, university observatories, and private foundations. Each piece of the work has to compete for funds on its merits. None of it requires large sums of money. No new telescopes have to be built; only new electronic cameras and new data-processing software are required. The various projects produce important discoveries at a price that funding agencies with limited resources can afford. The extreme of cheapness is achieved by Penny Sackett with her PLANET project. Her project so far has cost less than a hundred thousand dollars, much of it provided by the southern hemisphere observatories. The project works well because the observatories are on the internet. They can transmit data to each other and coordinate their activities from minute to minute. The style of the work is dominated by two factors, local enthusiasm and the internet. Penny Sackett happened to get acquainted with the southern hemisphere astronomers because she went to a meeting of the International Astronomical Union in Buenos Aires. She told them about gravitational microlensing, and they told her about their idle telescopes. Within a week she had the project organized. She took advantage of the opportunity offered by gravitational lensing to bring the four observatories together in a serious international enterprise, and they are enjoying for the first time the close working contacts which the internet makes possible.

Enterprises similar to Penny Sackett's project are playing an important role in the struggle to understand and control the outbreak of emerging virus diseases. Emerging viruses cause death and destruction to animals and crop-plants as well as to humans. Just as some of the most interesting celestial objects are best observed from remote southern observatories, some of the most interesting viruses are endemic in remote places in Africa. Many of the puzzles of virus epidemiology are studied by collaborative efforts, linking together isolated medical centers and laboratories in remote places. With modest funding, the internet brings together the local knowledge and dedication of African experts with the technological resources provided by big medical centers in Europe and Japan and the United States. In this way the style of Penny Sackett is translated into the language of medicine. More broadly, the style of the gravitational tomography enterprise as a whole provides a good model for the style of research in the epidemiological aspects of medicine. If you spend money sparingly and work with collaborators all over the world on the internet, you can make a few dollars go a long way.

*　　*　　*

The medical and biological communities may have something else to learn from the astronomers. Medical scientists and biologists now have an opportunity to start a revolution in the scientific understanding of viruses and cells. They could start a revolution if they could invent two essential tools that do not yet exist. One of these tools is the desktop DNA sequencer, using direct physical methods rather than wet chemistry to sequence individual molecules of DNA. If a physical sequencer could be made to work, stripping bases off a DNA molecule one at a time and running them through a mass-spectrometer at a rate of one every millisecond, a complete human genome could be sequenced in a month. The four species of bases that occur in DNA have four different masses, and there is no law of physics that says they could not be reliably identified in a mass-spectrometer at least as fast as this. Many other ways of using physical tools to sequence DNA have been suggested. Physicists and biologists at various places around the world are working on schemes using tunneling scanning microscopes or novel detectors of optical fluorescence. There is no reason why a physical sequencer should not sit on a desktop and sequence ten human genomes every year at a cost on the order of ten thousand dollars each. It could also sequence a viral genome in a few minutes, a bacterial genome in a few hours.

The second essential tool that biologists might invent is a desktop protein microscope that could determine the three-dimensional structure of protein molecules. Until now, the structures of proteins have mostly been determined by X-ray diffraction, a laborious and expensive process that requires the purification and crystallization of a macroscopic sample of the protein. Some structures have also been determined by nuclear magnetic resonance spectroscopy, but this technique only works for small proteins. Neither technique works well for membrane proteins, which are generally large and resistant to crystallization. Membrane proteins are the proteins that sit on the surfaces of cells, half inside and half outside the cell membrane. They are used by every cell to communicate with its neighbors and to defend itself against invasion. To understand human diseases and defense mechanisms in detail, it is essential to determine the structure of membrane proteins. A group in La Jolla recently succeeded in crystallizing a membrane protein, the T-cell receptor molecule that is crucial to the understanding of AIDS. They were able to determine its structure by X-ray diffraction, but this was a unique and heroic effort, taking a large team several years to accomplish.

Fifty years ago, Max Perutz and John Kendrew determined the structures of hemoglobin and myoglobin, the most important proteins in human blood and human muscles. Those were the first two protein structures to be found. Perutz and Kendrew each spent many years to analyze a single protein and won Nobel prizes for their efforts.

Since that time, only about five thousand protein structures have been determined in fifty years. Human cells contain about a hundred thousand different proteins. The lack of good tools for protein structure analysis is an even greater obstacle to progress in biology and medicine than the lack of good tools for sequencing of DNA. To understand in detail how viruses attack and invade cells, you need to determine structures of hundreds of thousands of proteins. Many of the most important are membrane proteins. Existing tools are inadequate to the task. One existing tool is the atomic force microscope, which works by scraping over the surface of solid objects with a sharp point and measuring the up-and-down movements of the point. The atomic force microscope has resolution sharp enough to see individual atoms but is unable to penetrate below the surface. Another existing tool is the magnetic resonance imager, which is used every day in hospitals to see tumors in human heads. The magnetic resonance imager penetrates well below the surface but cannot see structures smaller than about a millimeter in size. The desktop protein microscope would be a device combining the atomic resolution of the atomic force microscope with the penetration of a magnetic resonance imager. The conceptual design of such a device was published by John Sidles a few years ago. Sidles is a theoretical physicist working in the orthopedics department of the University of Washington medical school. He deals with human patients every day and invents imaginative tools in the evenings. Several groups of inventors at other places are attempting to develop other versions of high-resolution magnetic resonance imaging. If Sidles' dream of a magnetic resonance imaging microscope with atomic resolution could be made to work, it could precisely locate every nitrogen and hydrogen atom in a protein molecule or in a virus. Precise structures of protein molecules of all kinds could be determined cheaply and rapidly.

I am suggesting to medical scientists and biologists that they might make faster progress if they would follow the lead of the astronomers and invent their own tools. Two of the essential tools required for the next revolution in biology, the desktop sequencer and the desktop protein microscope, do not exist. Why should biologists wait for some physicist at the University of Washington to invent them and for some instrument company in Boston or Taiwan to sell them? Why should they not go ahead and develop the tools themselves? They have here an opportunity to be leaders of a revolution, not only in biology but in the practice of medicine. If they could create the tools to sequence DNA and determine protein structures a thousand times faster and cheaper than the competition, the world would beat a path to their door. They would be in a situation like John Randall in 1945, in possession of the tools and the prestige that would allow them to move ahead from this revolution to the next.

technology & social justice

In 1985 I gave a series of lectures in Scotland, and the lectures were published in a book with the title *Infinite in All Directions*. One of the chapters, "The Twenty-first Century," contained my guesses for the most important things to come. It included my list of the three most important technologies for the coming century: genetic engineering, artificial intelligence, and space travel. Now that the coming century is almost here, things look different. The twenty-first century has moved from the far to the near future. My predictions are less concerned with the long run and more concerned with the short run. In the short run, space travel is a joke. We look at the bewildered cosmonauts struggling to survive in the Mir space station. Obviously they are not going anywhere except, if they are lucky, down. Artificial intelligence is also doing poorly. Robots are not noticeably smarter today than they were fourteen years ago. When I sat down to revise my predictions for this book, I removed space travel and artificial intelligence from the list. The only item left from my old list is genetic engineering, which is forging ahead even faster than I expected. Everyone has heard about Dolly, the cloned sheep. Much more important than Dolly is the discovery that the basic patterns of genetic control of development are the same in yeast and fruit flies and mice and humans, so that we can learn from experiments with yeast and fruit flies and mice how human babies grow. Genetic engineering stays on the list of important things for us to worry about.

Meanwhile, during the last fourteen years, the internet and the World Wide Web have exploded. They have become the dominating technology in modern life. I did not foresee this fourteen years ago, and neither did Bill Gates. Now it is obvious that the internet must be on the list.

The third item on the new list was not so obvious. I chose the sun, because I see solar energy as a winner in the game of matching new technologies to human needs. I could be wrong about this. I have been wrong before when I made predictions, but it is better to be wrong than to be vague. The sun has at least a possibility of being a winner during the next century. The new list—the sun, the genome, and the internet—became the title of this book.

Although space travel and artificial intelligence disappeared from the title, they have not disappeared from the book. Artificial intelligence is inescapably involved in the growth of the internet. As the internet and the World Wide Web become more capable and more complicated, they develop a kind of intelligence, too dispersed and too agile for slow-moving human brains to comprehend in detail. My son, George, in his book *Darwin Among the Machines*, describes the long history of artificial intelligence, tracing it back three hundred years into the past. He sees it as a central theme in the evolution of human society, both past and future. I shall return briefly to the subject of artificial intelligence in the epilogue to this book.

There are two reasons why space is still an important part of our future, in spite of past disappointments and failures. First, even at present-day prices, space science is a cost-effective tool for exploring the universe, and space communication is a cost-effective tool for building bridges between people around the world. Space science and space communication will be flourishing during the first half of the twenty-first century. Second, radical reduction of the price of space operations is a real possibility. If prices can be radically reduced, the dream of enlarging the domain of life beyond this planet might become reality. But the new technologies that might spread life over the universe will come, if they come at all, during the second half of the century.

In this chapter I am examining technology from an earthbound point of view. I am looking for ways in which technology may contribute to social justice, to the alleviation of differences between rich and poor, to the preservation of the earth. The emphasis will be on technologies that could be developed within the next half-century, within the lifetimes of our children and grandchildren. In this near-term perspective, space communication may be important but space travel is irrelevant. In the next chapter I will take a longer view. Although space travel is today failing ludicrously to fulfill the promises of its promoters, it still might one day become cheap enough to be accessible to ordinary people. That day will not come soon. It will not be within the lifetimes of our children. But it might happen, before the end of the twenty-first century, that new technologies will be opening public highways into space for our great-grandchildren to travel.

*   *   *

It is easy to make a list of historical examples showing how technology has sometimes contributed to social justice. In the fourteenth century, the new technology of printing changed the face of Europe, bringing books and education out of the monasteries and spreading them far and wide among the people. Printing gave power to the Bible and led directly to the Protestant Reformation in northern Europe. One may question whether Luther's Germany and Shakespeare's England enjoyed social justice, but they were certainly closer to it than the medieval Germany and England out of which they grew. Luther and Shakespeare brought at least the idea of justice, if not the reality, to ordinary citizens outside the nobility and the priesthood. The Protestant ethic, which took root in Germany and England and Holland and Scandinavia with the help of printed books, carried with it a perpetual striving for social justice, even if the utopian visions were seldom achieved. Occasionally, in small communities, the utopian visions were achieved and social justice precariously established.

More recent technologies that contributed in a practical way to social justice were the technologies of public health, clean water supply, sewage treatment, vaccination, and antibiotics. These technologies could not protect the rich and powerful alone from contagion and sickness. They could only be effective in protecting the rich if they were also available to the poor. Even if the rich and powerful receive preferential treatment, as they usually do, the benefits of public health technology are felt to some extent by everybody. In countries where public health technologies are enforced by law, there is no large gap in life expectancy between rich and poor.

The technology of synthetic materials has also helped to erase differences between rich and poor. Throughout history until the nineteenth century, only the rich could afford to dress in brilliant colors, furs, and silk. Fine clothes were a badge of privilege and wealth. In the nineteenth century, chemical industry produced artificial dyestuffs. The twentieth century added artificial fur and silk, and many other synthetic fabrics, cheap enough for working-class women to afford. No longer can one tell a woman's social class by her clothes. It is a measure of social justice in modern societies that the children of rich families now dress down, imitating the style of the majority both in clothes and in behavior.

Household appliances are another technology with a tendency towards social justice. When I was a child in England in the 1920s, my mother employed four full-time servants: a cook and a housemaid, a nurse-maid and a gardener. We did not consider ourselves rich. My father was a schoolteacher. We were an average middle-class family. In those days an average middle-class family needed four servants, to do the hard manual work of cooking and cleaning and childcare and gardening. To do all this work, a whole class of people existed who spent their lives as domestic servants.

The professional and intellectual classes to which we belonged were riding on the backs of the servant class. Because of the servants, my mother had leisure to organize socially useful projects such as a club for teenage girls and a birth-control clinic. The birth-control clinic was undoubtedly a godsend to the women who came to it for instruction in the art of not having unwanted babies. It also gave my mother the satisfaction of bringing enlightenment to the lower classes.

But it did not in any way narrow the gulf between her and them. She always spoke of her birth-control clientele like a mistress speaking of servants.

My mother was a kind mistress and treated the servants well according to the standards of the time. But the servants knew their place. They knew that if they disobeyed orders or answered back, they would be out on the street. Our domestic arrangements were not altogether unlike the arrangements in the antebellum South of the United States. Like many of the household slaves in the old South, our servants seemed to be contented with their lot. We considered them to be our friends. Now, like the antebellum South, the servant class in England is gone with the wind. And the wind that blew away our servant class was not the ravaging invasion of Sherman's army, but the peaceful invasion of an army of electric stoves and gas heaters and vacuum cleaners and refrigerators and washing machines and driers and garbage disposals and freezers and microwave ovens and juicers and choppers and disposable diapers. The technology of household appliances made servants unnecessary, and at the same time the children of the servant class began to go to college and make the transition to the middle class. The transition was not painless, but it was less painful than a civil war. It was a big step on the road toward social justice. The middle class, which in my mother's time comprised only 10 percent of the population, grew rapidly until it included more than half the population.

I remember with great fondness the nursemaid Ethel who cared for me as a young child. She had left school, as girls of the servant class did in those days, at the age of fourteen. When my sister and I were safely in bed in the night nursery, we sometimes heard the putt-putt of a motorbike approaching the house, stopping, and then driving away into the night. That was Ethel's young man taking her out for the evening. The motorbike was the first harbinger of the approaching social revolution. It was the technology of upward mobility. After Ethel left us and married the young man, she had three daughters of her own and all of them went to college. One of her grandsons is a university professor.

Those are enough examples to show that technology can be helpful in the struggle for social justice. But in each case, as Edward Tenner tells us in his recent book, *Why Things Bite Back*, a step forward in technology tends to bring with it an unexpected step backward. A step forward in social justice for some people tends to bring with it a step backward for others. And it almost always happens that when an old privileged class of people is dispossessed and the blessings of wealth and power are spread more equally, the burdens of equalization fall disproportionately upon women. Thus, for example, when the spread of the technology of printing led to the Protestant Reformation that destroyed the wealth and power of the monasteries over much of Europe, both male and female orders were dispossessed, but the nuns lost more than the monks. Nuns in the old convents were in many ways more free than wives in the new Protestant communities. The old monastic society provided a refuge where women of outstanding ability, like Hildegard of Bingen, had access to higher education. Sheltered and supported by the monastic order, women could follow their vocations as scholars and artists. When the monasteries were dissolved, nuns had to find shelter in other people's homes, either as wives or as servants. The new secular society replaced the monasteries with colleges and universities. In the universities, male scholars could find shelter and security, but there was no place for women.

The technology of household appliances likewise brought a step backward to the stratum of society to which my mother belonged, the women of the old ser-vant-supported middle class. My mother would be considered by the standards of today a thoroughly lib-erated woman. Trained as a lawyer, she helped to write the Act of Parliament that opened the professions in England to women. With the help of her servants, she could take care of her husband and children without being confined to the home. She was free to pursue her interests outside the home, her girls' club and her birth-control clinic. But she was by no means the most liberated of the women in our family. I had a collection of aunts who were in various ways more liberated than my mother. All of them had husbands and most of them had children, but this did not stop them from being liberated. All of them were more adventurous than their husbands. My Aunt Margaret was trained as a nurse and rose to become matron, which meant that she was the managing administrator of a large hospi-tal. My Aunt Ruth was a figure skater of international repute, who kept an Olympic silver medal among her trophies. My Aunt Dulcibella was the first woman in England to receive an airplane pilot's license. She and her husband had an airplane that they used for travel-ing around in Africa. They loved Africa, and their lifestyle would have fit in well with the group of adven-turers that Michael Ondaatje describes in his novel *The English Patient*. Aunt Dulcibella was also a profession-al actress, and if she had been eighty years younger she might have had a starring role in the *English Patient* movie. We did not consider these aunts of ours to be unusual. It was considered normal at that time for mid-dle-class women to do something spectacular. My mother with her birth-control clinic was the quiet one, the least daring of the four.

Now consider what happened to the next genera-tion of middle-class women in England or in the Unit-ed States. Thirty years later, in the 1950s, the servants were gone and the electric appliances were taking their place. For wives and mothers of the middle class, this was a big step backward. Appliances do not cook the dinner, clean the house, do the shopping, and mind the baby. The middle-class women of the 1950s were far less liberated than their mothers had been. The liberation that my mother's generation achieved had to be fought for all over again. Even now in the 1990s, women are only partially liberated. To achieve even partial liberation, they have replaced the old domestic servants with day-care centers, cleaning ladies, and au pair girls imported from overseas. Electric appliances help, but they only do a small part of the job.

The Institute for Advanced Study where I have spent my working life is a peculiar institution with a small permanent faculty. The faculty is supposed to be representative of the most distinguished men and women in academic life. Unfortunately we have always found it difficult to appoint women to the faculty. The original faculty appointed in the 1930s contained one woman, the archeologist Hetty Goldman. I remember her vividly. She was a formidable lady, small in stature and large in spirit, who led excavations of ancient sites in Turkey, ruling over small armies of Turkish laborers with an iron hand. Her colleagues used to say that she was the equal of any two male archeologists. There was never the slightest doubt that she had the right stuff to be an Institute professor. She was a natural leader, in her own eyes and in ours.

She belonged to my mother's generation of liberated women. She grew up like my mother in a society of women with servants. When she retired in 1947, she was not replaced. It seemed that there was nobody of her preeminence in the next generation of women. For almost forty years, the Institute faculty was entirely male. In 1985 the sociologist Joan Scott became the second woman to join the faculty, and in 1997 the historian Patricia Crone became the third. The history of our faculty encapsulates the history of women's liberation, a glorious beginning in the 1920s, a great backsliding in the 1950s, a gradual recovery in the 1980s and 1990s. It is not altogether fanciful to blame the technology of household appliances for the backsliding. The advent of electrical appliances liberated the servants and shackled their mistresses.

* * *

I have mentioned four technologies that led to expansions of social justice. Although each of them had compensating negative effects, especially on women of the previously dominant classes, the overall effects of all of them were predominantly positive. It would be easy to find an equally impressive list of examples of technologies that had predominantly negative effects. One could begin with the technologies of gas chambers and nuclear weapons, useful for the convenient extermination of people to whom we do not wish to extend the benefits of social justice. But the more troubling examples are two of the technologies that are making the most rapid progress today, high-technology medicine and high-technology communication. Anyone who has had the misfortune to go, needing immediate attention, to the emergency room of a modern hospital, will be familiar with the ugly face that high-technology medicine presents to the patient. The long wait before anything happens, the filling out of forms, the repetitive answering of questions, the battery of routine chemical and physical tests carried out by masked technicians, and finally the abbreviated contact with the physician. The thing the patient needs the most, and the thing hardest to find, is personal attention.

Since personal attention has become the scarcest resource in high-tech medicine, it is inevitable that it should be distributed unequally. The majority of advanced countries have national health services that attempt, with varying degrees of success, to distribute medical attention fairly. In these countries, medical attention is theoretically available to everybody. This is what the ethic of social justice demands. But the escalating costs of medical attention make social justice more difficult to achieve. One way or another, as personal attention becomes scarcer, people of status tend to receive more of it and people without status tend to receive less. The national health services, in countries where they exist, make valiant efforts to preserve the ideal of social justice, and they are often successful. But the march of medical technology and the concomitant increase of costs are constantly eroding the ideal. In the United States, which has never had a national health service and does not pretend to distribute medical resources equally, the prospects for achieving social justice are far worse. In the United States, a medical system based on the ethic of the free market inevitably favors the rich over the poor, and the inequalities of medical treatment grow sharper as the costs increase.

I have seen in my own family a small example of the dilemma that the growth of high-tech medicine presents to physicians. One of my daughters is a cardiologist. For many years she worked in state-supported hospitals, taking care of patients as they flowed through the system, working brutally long hours and still having little time for personal contact with the patients. Her patients in the public hospitals were predominantly poor and uninsured. Many of them had AIDS or gunshot wounds in addition to cardiac problems. The public health system, such as it was, was designed to get these patients out of the hospital and back on the street as fast as possible. Last year my daughter was offered a job in a private cardiology practice, with far shorter hours, better pay and working conditions, and an expectation of long-continued care of her patients. She accepted the offer without much hesitation. She is much happier in her new job. Now for the first time she knows her patients as individuals. She can tailor their treatments to their individual histories and personalities. She feels that she is a better doctor. And her new job gave her the flexibility to take time off to have a baby. From almost every point of view, her jump into private practice was a wise move. Her only problem is a small twinge of conscience for having abandoned the poor to take care of the rich. In the private practice, her patients are not all rich, but they are all paying for the personal attention that she is now able to give them. She was forced to make a choice between social justice and professional satisfaction, and social justice lost. I do not blame her. In a socially just society, physicians would not be forced to make such choices.

Similar dilemmas, not as stark as the dilemmas of medical practice but equally important, exist in the world of high-technology computing and communications. Here too, there is a clash between the economic forces driving the technology and the needs of poor people. Access to personal computers and the internet is like medical insurance. Almost everybody needs it, but most poor people don't have it. The people who are wired, the people who browse the World Wide Web and conduct their daily lives and businesses on the internet, have tremendous economic and social advantages. Increasingly, jobs and business opportunities are offered through the internet, and access to the internet means access to well-paying jobs. People who are not wired are in danger of becoming the new servant class. The gulf between the wired and the unwired is wide and growing wider.

The computer and software industries are driven by two contradictory impulses. On the one hand, they sincerely wish to broaden their market by making computers accessible to everybody. On the other hand, they are forced by competitive pressures to upgrade their products constantly, increasing their power and speed, adding new features and new complications. The top end of the market drives the development of new products, and the new products remain out of reach of the poor. In the tug-of-war between broadening the market and pampering the top-end customer, the top-end customer usually wins.

The problem of unequal access to computers is only a small part of the problem of unequal opportunity in our society. Until the society is willing to attack the larger problems of inequality in housing and education and health care, attempts to provide equal access to computers cannot be totally successful. Nevertheless, in attacking the general problems of unequal opportunity, computer access may be a good place to start. One of the virtues of the new technology of the internet is that it has an inherent tendency to become global. The internet easily infiltrates through barriers of language, custom, and culture. No technical barrier stops it from becoming universally accessible. Providing equality of access to the internet is technically easier than providing equality of access to housing and health care. Universal access to the internet would not solve all of our social problems, but it would be a big step in the right direction. The internet could then become an important tool for alleviating other kinds of inequality. That is why the internet was given a place in the title of this book.

*   *   *

In the stories that I have been telling in this chapter, technology came first and ethics second. I have been describing historical processes in which technological changes occurred first, and then increases or decreases of social justice occurred as a consequence. This view of history is opposed to the view propounded by Max Weber in his famous book *The Protestant Ethic and the Spirit of Capitalism*. Weber argued that the Protestant ethic came first, the rise of capitalism and the technologies associated with capitalism second. Weber's view has become the prevailing view of modern historians. Weber said that ethics drove technology. I say that technology drives ethics.

I am not trying to prove Weber wrong. His historical vision remains profoundly true. It is true that the religious revolutions of the sixteenth century engendered an ethic of personal responsibility and restless inquiry, an ethic that encouraged the growth of capitalistic enterprise and technological innovation. It was no accident that Isaac Newton, the preeminent architect of modern science, was also a Protestant theologian. He took his theology as seriously as his science. It was no accident that King Henry VIII, the man who brought the Protestant revolution to England, also endowed the college where Newton lived and taught. Henry and Isaac were kindred spirits, both of them rebels against authority, both of them enemies of the Pope, both of them tyrants, both of them supreme egotists, both of them suspicious to the point of paranoia, both of them believers in the Protestant ethic, both of them in love with technology. Henry loved to build ships, and Isaac loved to build telescopes. It is true that ethics can drive technology. I am only saying that this is not the whole truth, that technology can also drive ethics, that the chain of causation works in both directions. The technology of printing helped to cause the rise of the Protestant ethic just as much as the Protestant ethic helped to cause the rise of navigation and astronomy.

I am not the first to take issue with Weber on this question. The historian Richard Tawney also studied the interrelationships of religion and capitalism and came to conclusions similar to mine. He held Weber in high esteem and contributed a foreword to the English translation of *The Protestant Ethic and the Spirit of Capitalism*. Here are the concluding sentences of Tawney's foreword: "It is instructive to trace, with Weber, the influence of religious ideas on economic development. It is not less important to grasp the effect of the economic arrangements accepted by an age on the opinion which it holds of the province of religion." Tawney's view is that technology influenced religion as strongly as religion influenced technology.

Since my view of history is closer to Tawney than to Weber, I now ask the question: how may we make ethics drive technology in such a way that the evil consequences are minimized and the good maximized? I shall argue that the chain of causation, from ethics to technology and back to ethics, leaves open a possibility of making technological progress and ethical progress run hand in hand. New technologies offer us real opportunities for making the world a happier place.

* * *

The new century will be a good time for new beginnings. Technology guided by ethics has the power to help the billions of poor people all over the earth. Too much of technology today is making toys for the rich. Ethics can push technology in a new direction, away from toys for the rich and towards necessities for the poor. The time is ripe for this to happen. The sun, the genome, and the internet are three revolutionary forces arriving with the new century. These forces are strong enough to reverse some of the worst evils of our time. One of the greatest evils is rural poverty. All over the earth, and especially in the poor countries to the south of us, millions of desperate people leave their villages and pour into overcrowded cities. There are now ten megacities in the world with populations twice as large as New York City. Soon there will be more. Mexico City is one of them. The increase of human population is one of the causes of the migration. The other cause is the poverty and lack of jobs in the villages. Both the population explosion and the poverty must be reversed if we are to have a decent future. Many experts on population say that if we can mitigate the poverty, the population will stabilize itself, as it has done in Europe and Japan.

I am not an expert on population, so I will say no more about that. I am saying that the poverty can be reduced by a combination of solar energy, genetic engineering, and the internet. And perhaps when the poverty stops increasing, the population will stop exploding.

I have seen with my own eyes what happens to a village when the economic basis of life collapses. My wife grew up in a village in East Germany that was under communist management. The communist regime took care of the village economy, selling the output of the farms to Russia at fixed prices which gave the farmers economic security. The village remained beautiful and on the whole pleasant to live in. Nothing much had changed in the village since 1910. One thing the communist regime did was to organize a zoo, with a collection of animals maintained by a few professionals with enthusiastic help from the local schoolchildren. The village was justly proud of its zoo. The zoo was subsidized by the regime so that it did not need to worry about being unprofitable. My wife and I visited the village under the old regime and found it very friendly.

Then came 1990 and the unification of Germany. Overnight the economy of the village was wrecked. The farmers could no longer farm because nobody would buy their product. Russia could not buy because the price had to be paid in West German marks. German consumers would not buy because the quality was not as good as the produce available in the supermarkets. The village farmers could not compete with the goods pouring in from France and Denmark. So the farmers were out of work. Most of the younger generation moved out of the village to compete for jobs in the cities. Most of the older generation remained. Many of them, both old and young, are still unemployed. The zoo, deprived of its subsidy, was closed. The sad exodus that we saw in the village of Westerhausen in 1991 is the same exodus that is happening in villages all over the world. Everywhere, the international market devalues the work of the village. The people of the village sink from poverty into destitution. Without work, the younger and more enterprising people move out.

In the nine years since the unification, Westerhausen has slowly been recovering. Recovery is possible because of the process of gentrification. The village is still a pleasant place to live if you have money. Wealthy people from the local towns move in and modernize the homes abandoned by the farmers. Cottages are demolished to make room for two-car garages. Ancient and narrow roads are widened to make room for two Mercedes cars to pass each other. The thousand-year-old church is being repaired and restored. The village will survive as a community of nature-lovers and commuters. Lying on the northern edge of the Harz mountains, it is close to the big cities of northern Germany and even closer to unspoiled mountain forests. Its permanent asset is natural beauty.

In 1997 my wife and I were back in the village. The change since we last visited in 1991 was startling. We stayed in the elegant new home of a friend who had been in my wife's class in the village elementary school fifty years earlier. The village now looks well maintained and prosperous. The recovery from the disaster of 1990 has been slow and difficult, but it has been steady. The government did two things to mitigate the harshness of the free market economy, to enable the native villagers who remained in the village to survive. Every homeowner can borrow money with low interest to modernize houses, and every farming cooperative can borrow money with low interest to modernize farms. As a result, many of the houses which were not bought by outsiders are being modernized, and the few farmers who remained as farmers are flourishing. The zoo has been revived. In addition, there are some new enterprises. A western immigrant has planted a large vineyard on a south-facing hillside and will soon be producing the first Westerhausen wines. My wife's family and many of her friends are remaining in the village. They gave us a warm and joyful welcome.

The probable future of Westerhausen can be seen in a thousand villages in England. The typical English village today is not primarily engaged in farming. The typical village remains beautiful and prosperous because of the process of gentrification. Wealthy homeowners pay large sums of money for the privilege of living under a thatched roof. The thatching of roofs is one of the few ancient village crafts that still survives. The thatchers are mostly young, highly skilled, and well paid. The farmers who remain are either gentleman amateurs who run small farms as a hobby, or well-educated professionals who run big farms as a business. The old population of peasant farmers, who used to live in the villages in poverty and squalor, disappeared long ago. Discreetly hidden in many of the villages are offices and factories engaged in high-tech industry. One of the head offices of IBM Europe is in the English village of Hursley, not far from where I was born.

In the villages of France, at least in the area that I know around Paris, the picture is similar. Wealth came to the villages because they have what wealthy people seek: peace and security and beauty.

What would it take to reverse the flow of people from villages to megacities all over the world, to stop the transformation of beautiful cities into unmanageable slums? I believe that the flow can be reversed by the same process of gentrification that is happening in Westerhausen. To make gentrification possible, the villages themselves must become a source of wealth. How can a godforsaken Mexican village become a source of wealth? Three facts can make it possible. First, solar energy is distributed equitably over the earth. Second, genetic engineering can make solar energy usable everywhere for the local creation of wealth. Third, the internet can provide people in every village with the information and skills they need to develop their talents. The sun, the genome, and the internet can work together to bring wealth to the villages of Mexico, just as the older technology of electricity and automobiles brought wealth to the villages of England. Each of the three new technologies has essential gifts to offer.

*   *   *

Solar energy is most abundant where it is most needed, in the countryside rather than in cities, in the tropical countries where most of the population lives rather than in temperate latitudes. Recently I got to know a young man called Bob Freling who runs a venture called SELF, the Solar Electric Light Fund. Freling is not a scientist. He is a linguist with a passion for languages. He is fluent in Spanish, Russian, French, Chinese, Portuguese, and Indonesian. He has the know-how to operate small enterprises in many different countries with different cultures and different ways of doing business. The idea of SELF is to bring electricity generated by sunlight to remote places that have no other way to get electricity.

A working solar energy system can make an enormous difference to the quality of life in a tropical village. Thirty or fifty watts of direct current is enough to run a couple of fluorescent lights, a radio, or a small black-and-white television for several hours every night. Each hut in a village can have its own system. No central generator, no power lines, no transformers are needed. Sunlight distributes power equally to each rooftop. Children can read and study at night in their homes. The village is in touch with the outside world.

SELF is one of the organizations dedicated to making this happen. It is a charitable foundation, but it does not give solar-energy systems to the villagers for free. The villagers pay market prices for the hardware. SELF gives them credit so they can spread their payments over four years. SELF also pays the people who train the villagers to install and operate and maintain the hardware. SELF now has village projects working in eleven countries, all in remote places unlikely to be reached in the near future by electric power lines. The most recent start was on the island of Guadalcanal, not far from the battlefield where one of the bloodiest campaigns of World War II was fought. Technology now brings light to the island instead of death and destruction. The village that is now electrified is still accessible only by canoe. The solar hardware is light and rugged enough to travel by canoe.

One day there will be a global internet carried by a network of low-altitude satellites linked by radio and laser communications. Every point on earth will be within range of one or more of the satellites all the time. But not every point on earth will be able to communicate with the internet. For places without electricity to power transmitters and receivers, the satellites overhead will be useless. The villages where SELF has supplied solar-energy systems will be hooked up to the net. Their neighbors all around will still be isolated.

Why do we need a charitable foundation to do this work? Why cannot the villagers do it by themselves? Unfortunately, the technology of solar energy is still too expensive for an average third-world village to afford. Even with the help provided by SELF, only villages with a substantial cash economy can afford it. The present cost of a minimal solar-energy installation is about five hundred dollars per household.

About half of this cost is due to the photovoltaic collector panels that convert sunlight into electricity. The other half is spent on accessories such as storage batteries and control circuitry. The cost of collector panels is now about five dollars per watt. This is the cost of commercially available units that are properly packaged to work outdoors in rough weather. Experimental units that are under development will be substantially cheaper. The prevailing belief among economists is that solar energy will not supplant kerosene and other fossil fuels on a massive scale until the price comes down below one dollar per watt. Nobody can tell when, if ever, the existing photovoltaic technology will become cheap enough to supply the world's needs. All that we know for sure is that there is enough sunlight to supply our needs many times over, if we can find a way to use it.

Meanwhile, other ways of using sunlight are making slower progress. The traditional ways to use sunlight are to grow crops for food and trees for fuel. Even today, with the forests rapidly disappearing, firewood is still a major source of energy for the rural population of India. Less destructive ways of converting sunlight into fuel exist. Plant and animal wastes can be digested in tanks to make methane gas. Sugar cane can be fermented to make alcohol. These technologies are being used successfully in many places. Progress is slow because coal and oil and natural gas are still too cheap. It is possible that energy crops, either trees or other plants grown for conversion into fuel, will one day be a major source of energy for the world. But this will not happen until the cost of growing and processing energy crops falls far below its present level.

The SELF projects in remote regions of Asia and Africa and Australasia are cost effective because the usual alternative source of energy for the villagers is kerosene, and kerosene in remote areas is expensive. Kerosene has to be carried in cans on the backs of animals or people. Solar energy needs no carrying. But the SELF projects are limited to a small scale because that is all the villagers can afford.

Fifty watts per household is not enough to support a modern economy. For the people in a remote village to become a part of the modern world, they need electricity in far greater quantity. They need kilowatts per household rather than watts. They need refrigerators and power tools and electrical machinery. If solar energy is to satisfy their needs, it must be available on a massive scale.

The flux of solar energy incident upon the earth is enormous compared with all other energy resources. Each square mile in the tropics receives about a thousand megawatts, averaged over day and night. This quantity of energy would be ample to support a dense population with all modern conveniences. Solar energy has not yet been used on a large scale for one simple reason: it is too expensive. The country that has used solar energy on the largest scale is Brazil, where sugar was grown as an energy crop to make alcohol to be used as a substitute for gasoline in cars and trucks. Brazil protected and subsidized the local alcohol industry. The experiment was technically successful, but the cost was high. Brazil has now reverted to free market policies, and the experiment is at an end. What the world needs is not high-cost subsidized solar energy but solar energy cheap enough to compete with oil.

Solar energy is expensive today because it has to be collected from large areas and we do not yet have a technology that covers large areas cheaply. One of the virtues of solar energy is the fact that it can be collected in many ways. It is adaptable to local conditions.

The two main tools for collecting it are photoelectric panels producing electricity and energy crops producing fuel. With the exception of small units such as those used in the SELF projects, energy crops are the method of choice for farmland and forests, while photoelectric collection is the method of choice for deserts. Each method has its advantages and disadvantages. Photoelectric systems have high efficiency, typically between 10 and 15 percent, but are expensive to deploy and maintain. Energy crops have low efficiency, typically around 1 percent, and are expensive and messy to harvest. The electricity produced by photoelectric systems is intermittent and cannot be cheaply converted into storable form.

Fuels produced from energy crops are storable. To make solar energy cheap, we need a technology that combines the advantages of photovoltaic and biological systems. Two technical advances would make this possible. First, crop plants could be developed that convert sunlight to fuel with efficiency comparable to photovoltaic collectors, in the range of 10 percent. This would reduce the costs of land and harvesting by a large factor. Second, crop plants could be developed that do not need to be harvested at all. An energy crop could be a permanent forest of trees that convert sunlight to liquid fuel and deliver the fuel directly through their roots to a network of underground pipelines. If these two advances could be combined, we would have a supply of solar energy that was cheap, abundant, and environmentally benign.

Our hopes for radical decrease of costs and increase of efficiency of energy crops must rest on the genome. Traditional farming has always been based on genetic engineering. Every major crop plant and farm animal has been genetically engineered by selective breeding until it barely resembles the wild species from which it originated. Genetic engineering as the basis of the world economy is nothing new. What is new is the speed of development. Traditional genetic engineering took centuries or millennia to produce the improved plants and animals that fed the world until a hundred years ago. Modern genetic engineering, based on detailed understanding of the genome, will be able to make radical improvements within a few years. That is why I look to the genome, together with the sun and the internet, as tools with which to build a brighter future for mankind.

The energy supply system of the future might be a large forest, with the species of trees varying from place to place to suit the local climate and topography. We may hope that substantial parts of the forest would be nature reserves, closed to human settlement and populated with wildlife so as to preserve the diversity of natural ecologies. But the greater part could be open to human settlement, teeming with towns and villages under the trees. Landowners outside the nature reserves could be given a free choice, whether to grow trees for energy or not. If the trees converted sunlight into fuel with 10 percent efficiency, landowners could sell the fuel for ten thousand dollars per year per acre and easily undercut the present price of gasoline. Owners of farmland and city lots would have a strong economic incentive to grow trees. Even without such an incentive, towns where wealthy people live are usually full of trees.

People like to live among trees. If the trees were also generating fuel from sunlight, the appearance of the towns would not be greatly altered. The future energy plantation need not be a monotonous expanse of trees of a single species planted in uniform rows. It could be as varied and as spontaneous as a natural woodland, interspersed with open spaces and houses and towns and factories and lakes.

To make this dream of a future landscape come true, the essential tool is genetic engineering. At present large sums of money are being spent on sequencing the human genome. The human genome project does not contribute directly to the engineering of trees. But alongside the human genome, many other genomes are being sequenced: bacteria and yeast and worms and fruit flies. For advancing the art of genetic engineering, the genomes of simpler organisms are more useful than the human genome. Before long we shall have sequenced the genomes of the major crop plants, wheat and maize and rice, and after that will come trees. Within a few decades we shall have achieved a deep understanding of the genome, an understanding that will allow us to breed trees that will turn sunlight into fuel and still preserve the diversity that makes natural forests beautiful.

While we are genetically engineering trees to use sunlight efficiently to make fuel, we shall also be breeding trees that use sunlight to make other useful products, such as silicon chips for computers and silicon film for photovoltaic collectors. Economic forces will then move industries from cities to the country. Mining and manufacturing could be economically based on locally available solar energy, with genetically engineered creatures consuming and recycling the waste products. It might even become possible to build roads and buildings biologically, breeding little polyps to lay down durable structures on land in the same way as their cousins build coral reefs in the ocean.

The third and most important of the triad of new

technologies is the internet. The internet is essential, to enable businesses and farms in remote places to function as part of the modern global economy. The internet will allow people in remote places to make business deals, to buy and sell, to keep in touch with their friends, to continue their education, to follow their hobbies and avocations, with full knowledge of what is going on in the rest of the world. This will not be the internet of today, accessible only to computer-literate people in rich countries and to the wealthy elite in poor countries. It will be a truly global internet, using a network of satellites in space for communication with places that fiber optics cannot reach, and connected to local networks in every village. The new internet will end the cultural isolation of poor countries and poor people.

There are two technical problems that must be solved to make an internet accessible to almost everybody. There is the problem of large-scale architecture and the problem of the last mile. Large-scale architecture means choosing the most efficient combination of land-lines and satellite links to cover every corner of the globe. The Teledesic system of satellite communication now under development is intended to be a partial answer to this problem. The Teledesic system has 288 satellites in a dense network of low orbits, allowing any two points on the globe to be connected with minimum delay. If the Teledesic system fails or is not deployed, some other system will be designed to do the same job. The problem of the last mile is more difficult. This is the problem of connecting individual homes and families, wherever they happen to be, to the nearest internet terminal. The problem of the last mile has to be solved piecemeal, with methods depending on the local geography and the local culture.

An ingenious method of solving the last mile problem in urban American neighborhoods has been introduced recently by Paul Baran, the original inventor of the internet. Baran's system is called Ricochet and consists of a multitude of small wireless transmitters and receivers. Each user has a modem that communicates by radio with the local network. The feature that makes the system practical is the ability of the transmitters to switch their frequencies constantly so as not to interfere with one another. The system is flexible and cheap, avoiding the large expense of laying cables from the internet terminal to every apartment in every house. It works well in the environment of urban America. It remains to be seen whether it is flexible and cheap enough to work well in the environment of a Mexican village or a Peruvian barrio. An essential prerequisite for access to the internet is access to a reliable local supply of electricity. Villagers must have electricity first, before they can have modems.

Suppose that we can solve the technical problems of cheap solar energy, genetic engineering of industrial crop plants, and universal access to the internet. What then will follow? The solution of these three problems could bring about a world-wide social revolution, similar to the revolution that we have seen in the villages of England and Germany. Cheap solar energy and genetic engineering will provide the basis for primary industries in the countryside, modernized farming and mining and manufacturing. After that, the vast variety of secondary and tertiary economic activities that use the internet for their coordination, food processing and publishing and education and entertainment and health care, will follow the primary industries as they move from overgrown cities to country towns and villages. As soon as the villages become rich, they will attract people and wealth back from the cities.

I am not suggesting that in the brave new world of the future everyone will be compelled to live in villages. Many of us will always prefer to live in large cities or in towns of moderate size. I am suggesting only that people should be free to choose. When wealth has moved back to the villages, people who live there will no longer be compelled by economic necessity to move out, and people who live in megacities will no longer be compelled by economic necessity to stay there. Many of us who have the freedom to choose, like successful stockbrokers and business executives in England and Germany, will choose to live in villages.

This is my dream, that solar energy and genetic engineering and the internet will work together to create a socially just world, in which every Mexican village becomes as wealthy as Princeton. Of course that is only a dream. Inequalities will persist and poverty will not disappear. But I see a hope that the world will move far and fast along the road that I have been describing. We need to apply a strong ethical push to add force to the technological pull. Ethics must guide technology in the direction of social justice. Let us help to push the world in that direction as hard as we can. It does no harm to hope.

The High Road

In 1985 I was visiting the Johnson Space Flight Center in Houston and climbing around in the space shuttle that they keep there for visitors to examine. That was before the Challenger disaster, when the shuttle was advertised as a safe ride for congressmen and schoolteachers. What impressed me about the shuttle was the immense quantity of stuff that they had on board for the care and comfort of human passengers. It felt more like a hotel or a hospital than a rocketship. I made rough calculations of how many tons of stuff they were flying to keep seven passengers alive and well for a couple of weeks. I was thinking, why don't they rip out all this stuff and fly the shuttle by remote control from the ground? The shuttle take-off and flight into orbit were already ground-controlled. Only the landing was done by a human pilot. It would be easy to use remote control for the landing too.

At that time, most of the shuttle missions were carrying unmanned satellites into orbit for various purposes, some of them scientific, some of them commercial, and some of them military. These launching jobs could just as well have been done automatically. Only a few of the shuttle missions really needed to have people on board, to do experiments or to repair the Hubble space telescope. It would have made sense to reserve two shuttle ships with all their hotel equipment for missions in which people were essential, and to use the other two shuttles for satellite-launching jobs with the passenger accommodations ripped out. The owners of railroad companies discovered long ago that it makes sense to use separate trains for passengers and freight. Passenger trains are expensive and passengers are fussy. Freight trains are cheaper and carry many more tons of payload. If we had run the shuttle program like a railroad, we would have saved a huge amount of money. The freight-only version of the shuttle could have carried bigger payloads for less money than the passenger version, without risking any lives. Unfortunately the authorities at Houston did not think that this was a good idea. Their whole existence was centered on the training of astronauts and the operation of manned missions.

After failing to eviscerate the shuttle, I wandered into the museum of the Johnson Space Flight Center, where there is a collection of rocks that astronauts brought back from the moon. Many of the moon rocks have been given or lent to scientists who want to examine them at other places, but a large number remain in Houston.

Scientists who are interested in moon rocks are usually also interested in meteorites. The tools that they use for analyzing meteorites are also good for analyzing moon rocks. So the scientists at Houston have organized expeditions to collect meteorites that have accumulated on the ice in Antarctica. Their meteorites sit in glass cases next to the moon rocks. Among their collection of meteorites were two that came from Mars. They also had in their collection the meteorite ALH84001, which afterwards became famous. That one was collected by Roberta Score in Antarctica in 1984 and brought back to Houston, but it was not identified as a Mars rock when I visited the museum. David Mittelfehldt identified it three years later.

The Mars rocks look different from the other meteorites. They are lighter in color and sandier in texture. But the difference in appearance does not prove that they came from a different place in the sky. We know for sure that they came from Mars because one of them contains tiny bubbles full of gas that has exactly the same composition as the atmosphere of Mars. The composition of the Mars atmosphere was measured by the instruments that were landed on Mars by the Viking missions in 1976. It is quite different from the atmosphere of earth. Gas with the composition found in the Mars rocks could not have come from anywhere except Mars. The Mars rocks must have been splashed off the surface of Mars by a big impact when an asteroid or a comet collided with the planet. They then orbited around the sun for a few million years until they happened to land on Earth. Besides the two that I saw in Houston, there are a dozen more in various other museums. It is surprising that so many of them survived the initial impact on Mars and the journey to Earth and also had the good luck to be picked up by meteorite hunters. Since so many have been found in a few years on Earth, there must be millions more Mars rocks floating in orbit around the sun, waiting to be discovered.

It seemed like a miracle. Here I was in the museum

in Houston, twelve inches away from a piece of Mars, with only a thin pane of glass to stop me from grabbing hold of it. In those days NASA was talking seriously about grandiose missions to Mars, costing many billions of dollars. One of the reasons for going to Mars was to return samples of rock to earth for scientists to analyze. And here were samples of Mars rock already in Houston, provided by nature free of charge. I found it odd that nobody seemed to be studying them. So far as I could tell, nobody in Houston seemed to be excited about the Mars rocks, except for me. I stood and gazed at the rocks for a long time. Nobody else came to look at them. I remarked to some NASA people that they might usefully spend some time studying the Mars rocks they already had, instead of planning billion-dollar missions to collect more. At that time the administrators in Houston did not seem to be interested in anything that did not cost billions of dollars.

Things have changed since then. Now NASA is interested in cheap missions, and many more scientists are interested in the Mars rocks. Recently, some of the Mars rocks were examined more thoroughly than ever before. The famous rock ALH84001, which was sitting in the museum in Houston when I visited but was not yet identified as a Mars rock, contains chemical traces that might be interpreted as evidence of ancient life on Mars. Two groups of scientists found little wormlike structures that might be relics of ancient microbes.

The evidence that these traces have anything to do with biology is highly dubious. On the basis of this evidence we cannot say that life must have existed on Mars. The discovery of these traces is important for two other reasons. First, if we are seriously interested in finding evidence for life on Mars, we now know that the Mars rocks on Earth are the most convenient place to look for it. Instead of waiting for many years for an expensive sample-return mission to land on Mars and return a few small chips of rock to Earth, we can find a supply of bigger chips lying in Antarctica, where meteorites accumulate on the ice and are free for us to take home. Second, these rocks show that if life was established on Mars at any time in the past, it was possible for it to be transported to Earth intact, so that life on Earth might be descended from life on Mars. Joseph Kirschvink at the California Institute of Technology did a careful mineralogical study of the Mars rock ALH84001 and concluded that the temperature inside the rock could never have risen much above the boiling point of water. Below the surface, most of the rock remained cool. If there had been a living microbe or spore inside the rock, it might conceivably have survived the voyage.

In the first billion years after the solar system was formed, when Mars had a warm climate and abundant water, asteroid impacts were much more frequent than they are now. In those early times, Mars rocks were falling on Earth in great numbers, and many Earth rocks must also have been falling on Mars. We should not be surprised if we find that life, wherever it originated, spread rapidly from one planet to another. Whatever creatures we may find on Mars will probably be either our ancestors or our cousins.

*   *   *

Another place where life might now be flourishing is in a deep ocean on Jupiter's satellite Europa. Jupiter has four large satellites, discovered almost four hundred years ago by Galileo. Galileo named them Io, Europa, Ganymede, and Callisto, arranged in order of increasing distance from Jupiter. He was well versed in classical mythology and knew that Io, Europa, and Callisto were girlfriends of Jupiter and Ganymede was a boyfriend. The Galileo spacecraft now orbiting around Jupiter is sending back splendid pictures of the satellites. The new pictures of Europa show a smooth icy surface with many large cracks but few craters. It looks as if the ice were floating on a liquid ocean and being fractured from time to time by movements of the water underneath. The pictures are strikingly similar to some pictures of the ice that floats on the Arctic Ocean of Earth. It would not be surprising if Europa should have a warm ocean under the ice. Io is blazing hot, with active volcanoes on its surface, Ganymede has an icy surface like Europa but not so smooth, and Callisto looks like a solid ball of ice covered with ancient craters. The three inner satellites are heated internally by the tidal effects of the huge mass of Jupiter, but the internal heating falls off rapidly with distance from Jupiter. We should expect that, below the surface, Europa would be much cooler than Io and much warmer than Ganymede and Callisto. Since Io is hot enough to boil away all its water, and Callisto is cold enough to freeze solid, Europa is the most likely place to find a warm liquid ocean. Ganymede might also have a liquid ocean, but it would be covered by a much thicker layer of ice. Of all the worlds that we have explored beyond the Earth, Mars and Europa are today the most promising places to look for life.

To land a spacecraft on Europa, with the heavy equipment needed to penetrate the ice and explore the ocean directly, will be a formidable undertaking. A direct search for life in the Europa ocean would today be prohibitively expensive. But there is an easier way. Just as asteroid and comet impacts gave us an easier way to look for evidence of life on Mars, impacts on Europa give us an easier way to look for evidence of life on Europa. We know that impacts on Europa must be at least as frequent as impacts on Callisto, the more distant satellite of Jupiter that is visibly covered with craters. Every time there is a major impact on Europa, a vast quantity of water will be splashed from the ocean into the space around Jupiter. The water will partly evaporate and partly condense into snow. Any creatures living in the water not too close to the impact will have a chance of being splashed intact into space with the water and quickly freeze-dried. Therefore, the easy way to look for evidence of life in the Europa ocean is to look for freeze-dried fish. The fish might be lying stranded on the surface of Europa, or they might be orbiting around Jupiter. Jupiter already has a ring of space debris orbiting around it. It is likely that freeze-dried fish or any other garbage splashed out of the Europa ocean will accumulate in the ring. Bringing a spacecraft to visit and survey the Jupiter ring would be far less expensive than bringing a submarine to visit and survey the Europa ocean. We might find many unexpected surprises in the Jupiter ring, even if we do not find freeze-dried fish. We might find freeze-dried seaweed or a freeze-dried sea monster.

Freeze-dried fish orbiting around Jupiter may be a fanciful notion, but nature in the biological realm has a tendency to be fanciful. Nature is usually more imaginative than we are. Nobody in Europe ever imagined a bird of paradise or a duck-billed platypus before they were discovered by explorers. Even after the platypus was discovered and a specimen brought to London, several learned experts declared it to be a fake. Many of nature's most beautiful creations might be dismissed as wildly improbable if they were not known to exist. When we are exploring the universe and looking for evidence of life, we may either look for things that are probable but hard to detect, or we may look for things that are improbable but easy to detect. In deciding what to look for, detectability is at least as useful a criterion as probability. Primitive organisms such as bacteria and algae hidden underground may be more

probable, but freeze-dried fish in orbit are more detectable. To have the best chance of success, we should
keep our eyes open for all possibilities.

A similar logic suggests warm-blooded plants as a
reasonable target for a search for life on the surface of
Mars. Any form of life that survived on Mars from the
early warm and wet era to the present cold and dry era
had a choice of two alternatives. Either it adopted an
entirely subterranean lifestyle, retreating deep underground to places where liquid water can be found, or it
remained on the surface and learned to protect itself
against cold and dryness by growing around itself an
insulating greenhouse, maintaining inside the greenhouse a warm and moist environment. The first alternative is more probable, but much more difficult to
detect. Organisms living deep underground without
making use of sunlight would probably be microscop-
ic, like the bacteria that live deep in the earth. To find
such organisms would require deep drilling and heavy
machinery. The second alternative is less probable but
more detectable.

Many species of terrestrial plants, including the
skunk cabbage that sprouts in our Princeton woods in
February, are warm-blooded to a limited extent. The
skunk cabbage maintains a warm temperature for
about two weeks in the part of its anatomy known as
the spadix, which contains the hidden flowers with
their male and female structures. According to folk-
lore, the spadix is warm enough to melt snow around
it. The evolutionary advantage of warm-bloodedness
to the plant is probably to attract small beetles or other
insects that linger in the spadix and pollinate the flowers. The warm temperature is maintained by rapid
metabolism of starch stored inside the spadix. The
spadix is not a greenhouse, and the supply of starch is
not sufficient to maintain a warm temperature all year

round. No terrestrial plants are able to stay warm through an Arctic winter. On Earth, polar bears flourish in colder climates than trees. It seems to be an accident of history that warm-blooded animals evolved on earth to colonize cold climates while warm-blooded plants did not. On Mars, because of the more severe pressure of natural selection, plants might have been pushed to more drastic adaptations.

Plants that learned to grow greenhouses could keep warm by the light from a distant sun and conserve the oxygen that they produce by photosynthesis. Plants could grow greenhouses just as turtles grow shells and polar bears grow fur. Colonies of such plants could build large structures, like the polyps that build coral reefs in tropical seas. The greenhouse would consist of a thick skin providing thermal insulation, with small transparent windows to admit sunlight. Outside the skin would be an array of simple lenses or mirrors focusing sunlight through the windows into the interior. The windows would have to be small, to limit loss of heat by outward radiation. The plant would also need deep roots to tap water and nutrients from warmer layers underground. The adjective "warm-blooded" does not mean that the plant must have a circulatory system or a precise temperature control. It means only that the plant is able to keep its internal temperature within the normal range of a cool greenhouse. Inside the greenhouse, the plant could grow leaves and flowers in an oxygen-containing habitat where aerobic microbes and animals may also live. Groups of greenhouses could grow together to form extended habitats for other species of plants and animals. An attendant community of microbes and fungi may help the plants to extract nutrients from the local minerals, ice, and soil. Pores in the outer skin of the greenhouse may open to admit carbon dioxide from the atmosphere outside, with miniature airlocks and cold traps to reduce losses of oxygen and water to a minimum.

If warm-blooded plants exist on Mars, they may or may not be easy to see. We cannot predict whether they would stand out from their surroundings in a visual or photographic survey. Two clues to their presence would almost certainly be detectable: leakage of heat and leakage of oxygen. Neither thermal insulation nor atmospheric containment is likely to be perfect. If we look for heat-radiation from anomalously warm patches on the Martian surface at night, or for anomalous local traces of oxygen in the atmosphere in daytime, we will find the places where warm-blooded plants might be hiding. Finding warm-blooded plants living wild on Mars or elsewhere in the solar system is only a remote possibility. It is much more likely that we shall find Mars sterile, or inhabited only by subterranean microbes. In that case, warm-blooded plants may be important in a different way, not as a goal for science but as a tool for human settlement. I now leave the subject of science and talk about human space travel and settlement. Human settlement of the solar system is not primarily a scientific enterprise. It is driven by motivations that go far beyond science.

*   *   *

The space shuttle program ran into deep trouble, even before the Challenger accident, because it was based on a confusion of aims. It was trying to serve as a practical launch system for scientific and commercial and military missions and also to open the way to human adventure in space. As we saw when we climbed into the shuttle at Houston, the two aims were never compatible. The shuttle was the result of a political compromise between people who wanted a reliable freight service into space and people who wanted to keep alive the tradition of the manned Apollo missions to the moon. No single vehicle could do both jobs well. The shuttle tried to do both jobs but was too expensive for the firſt and too limited in its performance for the second. The designers of the shuttle should not be blamed for its deficiencies. The confusion of aims that bedeviled the shuttle was imposed on the designers by political decision makers. The confusion of aims of the shuttle was a consequence of a much deeper confusion of aims of the U.S. space program as a whole.

From its beginnings in the 1950s, the U.S. space program had two diſtinĊt aims: the praĊtical maſtery of space for military and scientific purposes, and the idealiſtic vision of human adventure beyond the earth. Both aims were shared and loudly proclaimed by the Soviet Union. The rivalry with the Soviet Union provided the driving force that impelled the United States to pursue both aims vigorously for fifteen years. The firſt aim was embodied in the multitude of unmanned missions of the 1960s, ranging from the Atlas launchers and the Corona spy satellites to the Mariner explorations of the planets. The second aim was embodied in the Apollo moon landings, which were presented to the world on television with all the fanfare of an international sporting event. After the triumphs of Apollo, the incentive provided by Soviet competition faded. It became clear that the United States would no longer support two grand-scale programs pursuing the two aims separately. If both aims were to be maintained, a single major program would have to take care of both. The decision was made to abandon the Apollo program but to maintain the aims of Apollo in a compromise program that would also cover the praĊtical aims of freight transportation. The compromise program was the shuttle, saddled from its birth with the incompatible aims inherited from Atlas and Apollo.

The confusion of aims that affliĊted the U.S. space

program from the beginning was in essence a confusion of timescales. The practical aims of scientific and military activities in space made sense on a timescale of ten years. The basic technology of unmanned space missions took only ten years to develop. On the other hand, the aim of opening the skies to human exploration and adventure made sense on a timescale of a hundred years. It will probably take about a hundred years to develop the technologies that will allow significant numbers of human explorers to roam the skies at a price that earthbound citizens are willing to pay. The Apollo missions gave a false start to human exploration because they were tied to a ten-year timescale. The missions were far too costly to be sustained, and at the end of the ten-year program they came to a stop with no way ahead.

If it had been made clear from the beginning that manned exploration would be a hundred-year program with a stable and affordable budget, the confusion of aims would have been avoided. We might have developed a light two-passenger spacecraft, carrying enough fuel to explore far from the Earth, instead of building a shuttle that cannot go beyond low Earth orbit. A few explorers might now have been living on the moon, learning how to survive there permanently, using only the local resources.

We are now at the beginning of a revolution in space technology, when for the first time cheapness will be mandatory. Missions that are not cheap will not fly. This is bad news for space explorers in the short run, good news in the long run. The good news is that cheapness now has a chance. During the 1980s, missions to the planets were few and far between, because they became inordinately expensive. The fundamental reason why space science missions became expensive was the imbalance in funding between ground-based and space-based science. For thirty years, it was politically easier to obtain ten dollars for a space science mission than to obtain one dollar to do aſtronomy on the ground. Under these rules, ground-based aſtronomy became parsimonious while space science became extravagant.

The unfair competition injured both parties, ſtarving ground-based aſtronomy and spoiling space science. The injury to space science was greater. Ground-based aſtronomy flourished in spite of ſtarvation, while planetary missions almoſt ground to a halt in spite of big budgets. The rules are now changed in the direction of fair competition between ground and space. In recent years, planetary missions have been smaller and cheaper. In the future, all space science missions should be cheap. Once the barrier of high coſt is broken, missions will be more frequent and the pace of discovery will be faſter.

The twenty years that have elapsed since the birth of the shuttle have seen spectacular progress in the technologies of data processing, remote sensing and autonomous navigation. With these new technologies now available, almoſt all practical needs of science and commerce and national security are better served by unmanned missions than by the shuttle. In the future the two aims of the space program will be pursued separately. The shape of the future unmanned program pursuing practical aims is already visible, evolving from the exiſting program by incremental ſteps. Coſts of operations will be reduced by new technology and increasing autonomy of spacecraft. The future shape of the manned program pursuing idealiſtic aims is the great unknown. The shuttle is inadequate as a vehicle of human adventure. It resembles a Greyhound bus rather than a Landrover. Another spending spree like Apollo would be another dead end. It would be unsuſtainable, even if it were politically possible. Does the manned space program have a future? What are the requirements for a program that fulfills the long-term aims of human exploration and settlement beyond the earth at a cost that taxpayers will find reasonable? These are the questions that the promoters of manned missions must try to answer.

To make either unmanned or manned operations cheap, the first essential step is to eliminate the standing army of people who sit at mission control on the ground and take care of spacecraft day after day. Spacecraft and the instruments they carry must become completely autonomous. The new generation of unmanned missions has already achieved some degree of autonomy, with the result that a smaller number of people on the ground can take care of more missions.

The second step toward cheapness is more difficult. The second step is to find new ways to launch payloads from the surface of the earth into space. To make launch systems cheap, radical changes are needed. Three new launch systems, radically different from anything available today, are laser propulsion, ram accelerators, and slingatrons. If any of them can be made practical, they offer possibilities of reducing launch costs by a large factor.

*   *   *

Laser propulsion was invented and promoted by Arthur Kantrowitz, an expert in high-temperature gas dynamics. Thirty years ago, he was at the AVCO laboratory in Everett, Massachusetts, running a program aimed at practical applications of lasers and shock waves. He began small-scale experiments on laser propulsion.

He zapped objects made of various materials with pulsed lasers, burning off thin layers from their surfaces, and measured the recoil momentum delivered to the objects. He found that to transfer momentum efficiently it is best to use a double pulse. First you apply a short pulse to vaporize a thin layer. You wait a few microseconds to let the vapor expand a short distance away from the surface, and then you heat the vapor with a longer high-energy pulse that drives a shock wave into the vapor and applies a sustained pressure to the surface. The trick is to push hard on the surface without heating it too much. After the double pulse, you wait about a millisecond until the vapor has expanded out of the way. Then the process is repeated with another double pulse. For a full-scale laser-propulsion system you would need a high-power laser repeating the double pulse about a thousand times a second. The bottom surface of the spacecraft would be a flat porous plate, with a liquid propellant such as water forced through the pores. The water would form a thin layer on the surface between pulses and would be boiled away by the pulses while protecting the solid plate from melting. The flat plate with the tank of propellant above it would be a laser-propulsion motor. For many years, Arthur Kantrowitz tried to demonstrate that a flat-plate motor could generate enough thrust to lift itself against gravity. He never achieved this objective. With the lasers available to him, his model motors were not strong enough to lift their own weight off the ground.

There were various reasons why Kantrowitz never built a successful working model. First, he did not have the right kind of laser. The AVCO laboratory specialized in gas-dynamic lasers, which Kantrowitz himself had invented. These are carbon dioxide lasers, converting the energy of molecules in a supersonic jet into intense beams of infrared radiation. They are splendid for producing a steady continuous output at high power. They are not good for producing short pulses. The pulsed lasers that Kantrowitz had available were too feeble to deliver momentum efficiently to a motor. The second reason why the development of laser propulsion was stalled for twenty years was haphazard funding. I used to sit on a committee set up to advise ARPA (the Advanced Research Projects Agency of the Department of Defense) on the funding of laser-propulsion projects. ARPA was under pressure from the Defense Department, either to demonstrate the feasibility of laser propulsion quickly or to shut the program down.

The ARPA money was usually handed out for six months or a year at a time. The recipients of the money were told they must complete a "proof of principle" experiment during this brief time if they wanted their funding to continue. "Proof of principle" was the bureaucratic jargon phrase, meaning a successful model experiment. There was never enough time to do a careful experiment. Lasers had to be scrounged or borrowed from other programs, whether they were suitable for the experiment or not. Generally the money was spent before an adequate experiment could be done, and the contract was then terminated. After twenty years of spasmodic efforts, no convincing proof of principle experiment had been done. The basic reason why the program failed was that the Defense Department had no military requirement for laser propulsion. Laser propulsion should have been a civilian rather than a military venture. But NASA, which might have supported the program consistently as a long-range investment, was not interested.

The first successful flight of a laser-propulsion engine occurred in 1997. The flight was recorded on film, and the film was shown at the Thirteenth Space Manufacturing Conference at Princeton by Professor Leik Myrabo. The Space Manufacturing Conference is a biennial event organized by the Space Studies Institute, a private research institute founded by the late Gerard O'Neill to support commercial activities in space. Leik Myrabo is professor of electrical engineering at Rensselaer Polytechnic Institute. He has carried on the development of laser propulsion after Kantrowitz retired and has succeeded in making it work. We saw the proof of principle experiment in the film that he showed in Princeton, complete with the sound of a high-power pulsed laser firing ten times per second. The laser sounded like a machine gun. The engine is a graceful piece of light metal shaped like a mechanical fish, with a blunt nose, a reflecting ring around its waist, and a pointed tail. Myrabo calls it the Lightcraft Technology Demonstrator. It is six inches in diameter and weighs two ounces. It flew up along the beam of a ten-kilowatt laser belonging to the U.S. Air Force at White Sands, New Mexico. Each pulse of the laser was focused by the bottom surface of the Light-craft so as to create a high-temperature shock wave in the air beneath it. When the Lightcraft is flying in air, the air is the propellant and no additional liquid pro-pellant is needed. The repeated shock waves drove it upward with an acceleration of about two gees. The shock waves make intense flashes of light as well as sound but do no damage to the engine. The Lightcraft rose three feet into the air, not as high as the Wright brothers' first flight at Kitty Hawk, North Carolina, in 1903. [Leik Myrabo was back in Princeton in October 1998 with another film showing more recent flights of the Lightcraft. We saw it fly several times to a height of 75 feet. To go much higher than this will require a better laser.]

The era of cheap large-scale air transportation began fifty years after the Kitty Hawk flight. If we are lucky, the era of cheap space transportation might begin fifty years after the flight of the Lightcraft.

Now I switch from modest realities to glorious dreams. Here is a dream of a future laser-propulsion launch center. Imagine a big laser sitting on top of a mountain, the beam pointing up into the sky. Imagine a spacecraft, consisting of a flat tank full of water at the bottom and a compartment for payload and crew on top. The spacecraft sits in the back of a pickup truck. The truck drives up the road to the mountaintop; the spacecraft with the water tank is hoisted off the truck by a crane and suspended over the laser. The spacecraft might contain a couple of human passengers or some other fragile payload that cannot survive high acceleration. Water from the tank is forced through pores in a base-plate at the bottom, and the laser is switched on. With a blinding light emitted from the laser-heated water plasma, and a screaming sound with the pitch of a high soprano C emitted by the thousand pulses per second of exploding vapor, the laser motor begins to push the spacecraft upward. The spacecraft rides up the beam with an acceleration of three gees, reaching the velocity of escape from the earth in six minutes at a slant range of a thousand miles. The mass of the spacecraft at launch is two tons, the propellant is one ton of water contained in a light plastic tank with a flat steel bottom, and the final mass of the spacecraft with payload is one ton. The power of the laser beam is a thousand megawatts, converted into thrust with an overall efficiency of 15 percent.

At the end of six minutes of powered flight, the spacecraft is in an escape orbit and the laser is ready for the next launch. If the launcher is in use for 60 percent of the time, it can launch half a million spacecraft per year. The launch system is operated like a public toll road, available to anybody who arrives at the launcher with a spacecraft and enough money to pay the toll. The tolls cover the capital and operating costs of the launcher. The customers provide their own spacecraft and travel at their own risk.

This grand-scale laser launch system, with a one-ton payload and a low acceleration, is designed to carry human passengers. It would be wise to begin with a more modest system that one might call minilaser propulsion, carrying freight only, with an acceleration of thirty rather than three gees. The laser power is reduced from a thousand to a hundred megawatts, the payload is reduced from a ton to twenty pounds, the acceleration time is reduced from six minutes to thirty-six seconds, the energy-cost per pound remains unchanged, and the capital cost is reduced by a factor of ten. After a minilaser launch system has been put into operation and operated profitably, the risks and costs of a full-scale passenger-carrying system could be realistically assessed.

The key to cheap space launch is to have a high volume of traffic. The energy cost of a laser launch from earth into space is about five dollars per pound of payload. If the launcher is kept busy launching half a million tons of payload per year, collecting a toll of ten dollars per pound, the gross income is ten billion dollars a year, enough to cover capital repayment and maintenance in addition to energy costs. If the traffic is only a few launches per day, the toll will be several thousand dollars per pound, and the system will be an economic disaster like the space shuttle, a prestige project too expensive to be used for ordinary commercial activities. The worst folly would be to build a laser launch system before the demand exists to keep it active for a substantial fraction of the time.

These dreams of cheap laser launch systems will not come true until a reliable heavy-duty laser-propelled motor has been built and flown. The road from the two-ounce Lightcraft to a full-scale motor with a thrust of several tons will be a long one. To develop a new motor is always a difficult struggle, requiring large investments of money and human talent. This is a job that Myrabo and his students cannot do alone. A large aerospace company might one day decide to do it, with or without a little help and encouragement from NASA.

* * *

A totally different launch system, also radically new and potentially even cheaper than laser propulsion, is the ram accelerator, an inside-out ram-jet engine, invented and promoted by Abraham Hertzberg at the University of Washington in Seattle. In essence, it is only a very efficient version of a gas gun, accelerating a projectile inside a steel pipe.

The pipe is filled with a combustible mixture of gases. The projectile travels down the pipe, propelled by the steady pressure of the gas igniting behind it. The pipe is wider than the projectile, so that there is a gap filled with gas between the projectile and the pipe. The gas does not move with the projectile but mostly stays in place as the projectile passes by. The nose of the projectile is sharply pointed so that the bow shock is not strong enough to ignite the gas. The reflected shock at the stern is much stronger and causes the gas to ignite. I saw a toy model of the ram accelerator in action many years ago in the cellar of the engineering department at the University of Washington. The model was built and operated by two students, with Hertzberg telling them how to do it. The pipe was four inches in diameter and thirty feet long. The gas was a mixture of methane and air at twenty-five atmospheres pressure. The projectile was a pointed aluminum cone with fins to keep it in the center of the pipe. At the beginning of the run, the gas behind the projectile was ignited with a spark. For the rest of the run the gas burned smoothly and the projectile accelerated with a steady acceleration of thirty thousand gees. It came out of the end of the pipe at one and a half miles per second and was stopped in a tank filled with scraps of oriental rugs. When I visited the cellar and saw the operation, the model had already been fired five hundred times, and the inside of the pipe did not show a single scratch. The projectile automatically centered itself in the pipe as it flew along. Thirty thousand gees would not be healthy for human passengers, but would be tolerable for bulk freight, or for electronic machinery if carefully packed and loaded.

Hertzberg proposes to extrapolate his toy accelerator into a full-scale space launch system. The launcher would be built on the side of a mountain, pointing up into space. To escape from the earth, the payload needs to reach a speed of eight miles per second, five times the speed of the toy model. To reach five times the speed with the same acceleration, the pipe would be 750 instead of 30 feet long.

All things considered, 750 feet is not a very long distance. It is much shorter than an average airport runway. The payload would be long and slender, so that it could force its way through the atmosphere without losing a significant fraction of its momentum to drag forces. The main difficulty in extending the accelerator into the hypersonic domain is that the mixture of gases in the pipe must change as the speed increases from one end to the other. The speed of sound in the gas must increase with the speed of the projectile. The gas mixture must change by stages from methane and air to hydrogen and oxygen. Hertzberg envisages five stages, all at the same pressure, with thin diaphragms separating them in the pipe. As the projectile flies along the pipe, the diaphragms rupture without impeding its passage. The pattern of gas flow changes from one stage to another, but the acceleration remains roughly constant.

The ram accelerator is a low-tech device, in principle much cheaper than the high-tech laser-propulsion launch system. The energy costs of the ram accelerator are lower, and the capital costs are likely to be much lower. But the lower costs are accompanied by severe inconveniences. The ram accelerator can only accept payloads capable of withstanding high stress and fitting into long thin containers. The laser system can accept any payload that does not exceed its weight and size limits. In the long run, both systems will probably be economic, the ram accelerator for bulk freight and the laser launcher for general cargo.

The third possibility for a cheap space launch system is the slingatron, recently invented by Derek Tidman at a company called Datassociates in Virginia. Tidman has built a tabletop demonstration model that makes the concept easy to understand. The tabletop model is a hollow ring made of one-inch steel pipe, supported at six points by connections attached to six small horizontal wheels that are driven by electric motors. Inside the pipe is a small steel ball free to roll around the ring. The diameter of the ring is three feet, so that it fits easily onto a table. The motors cause the six support points of the ring to move together around small horizontal circles. The ring is forced into a circular motion, its size and orientation remaining unchanged. The position of the ball in the ring is monitored by a magnetic metal detector, and the motors driving the ring are programmed so that the motion of the support points around their small circles is always just a quarter of a cycle ahead of the motion of the ball. The motion of the ring and the motion of the ball speed up together. The ring is always moving inward in the direction toward its center at the place where the ball touches it. As a result, the ring pumps energy into the ball, and the ball is continuously accelerated, reaching a speed of 300 feet per second in a few seconds. The energy of the ring remains always small compared with the energy of the ball. For the system to work it is essential that the friction between ball and ring should be small. The frictional drag force must be less than the accelerating force exerted by the ring on the ball. This condition is easy to satisfy in the tabletop model, with a rolling ball and a small ring.

Tidman's tabletop slingatron is a splendid toy for children and grown-ups to play with. To go from the tabletop model to a full-scale space launch system is a big jump. A space launch slingatron would need to pump energy into a payload whirling around inside a ring until the payload reached a speed of seven or eight miles per second. The space inside the ring must be a fairly good vacuum. The ring radius must be hundreds or thousands of feet, to keep the centrifugal force between the ring and the payload within tolerable limits. The design that looks most promising uses a spiral track rather than a circular ring. Then the track is moved at a constant speed, while the payload moves outward along the spiral as it accelerates.

The frictional drag must be even smaller in the full-scale slingatron than it is in the model. The requirement of almost zero friction is the main difficulty of the slingatron. The payload to be accelerated must slide around the track with less friction than a skater on ice. Tidman has worked out several different ways to deal with the problem of friction. One way is to use a gas bearing, so that there is no surface contact between the payload and the track. A film of high-pressure gas, squeezed into the bearing layer between the underside of the payload and the track, acts as a lubricant. The operation of gas bearings at high speeds is a difficult and demanding technology. For a space launcher, there must also be a smooth transfer of the payload from the track to a launching ramp pointing upward into space, and an airlock allowing the payload to emerge from vacuum into air. The payload must have a low enough aerodynamic drag to push its way through the atmosphere into space without losing too much momentum. Many challenging problems of engineering design still need to be addressed.

The slingatron payload would have to withstand a peak acceleration of about a thousand gees, far larger than the laser launch acceleration but smaller than the ram accelerator acceleration. Until the design of a full-scale slingatron is worked out in detail, any estimate of its cost is a pure guess. It is possible that the slingatron may find a niche of economic viability, between the gentle but expensive laser launcher and the brutal but cheap ram accelerator.

The three launch systems that I have discussed have one feature in common that will be of great economic importance if any of them turns out to be practical. All of them launch payloads into escape orbits or high earth orbits as cheaply as into low earth orbits. In this respect they differ radically from chemical rocket launchers. With chemical rockets, the cost of putting payload into high orbit or escape is about ten times the cost of putting it into low orbit. If we ever have high-volume traffic and industry in space, we will need warehouses and refueling stations and factories in space too. If the public highway launch systems are going directly to high earth orbit, that is where the hub facilities of space industry and transport will be. Low earth orbit will remain a good place for observing the Earth and for rapid point-to-point communication networks. For access to the rest of the solar system, high earth orbit is far better. It is fortunate that high earth orbit is also a good place to put astronomical instruments, so they can see the whole sky without the Earth getting in the way. The successors of the Hubble space telescope will be orbiting far away from the Earth, perhaps even far away from the sun, where the views are unobstructed and the skies are clearer.

The coming era of cheap space operations will begin with unmanned missions. Cheap manned missions will come later, after unmanned missions have tested and exercised the new technologies of launch and operation. Cheap unmanned missions only require new engineering. Cheap manned missions will require new biotechnology. For a manned mission, the chief problem is not getting there but learning how to survive after you got there. To survive and make yourself a home away from Earth is a problem of biology rather than a problem of engineering.

It is easy to predict that cheap missions, both unmanned and manned, will ultimately be possible. There is no law of physics or biology that forbids cheap travel and settlement all over the solar system and beyond. But it is impossible to predict how long this will take. Predictions of the dates of future achievements are notoriously fallible. My guess is that the next fifty years will be the era of cheap unmanned missions, and that the era of cheap manned missions will start some time late in the twenty-first century. The time these things take will depend on unforeseeable accidents of history and politics. My date for the beginning of cheap manned exploration and settlement is based on a historical analogy. From Columbus's first voyage across the Atlantic to the settlement of the pilgrims in Massachusetts was 128 years. So I am guessing that in 2085, 128 years after the launch of the first Sputnik, the emigration of pilgrims from the earth may be beginning.

*   *   *

The main lesson that I draw from the history of past space activities is that we must clearly separate short-term from long-term aims. The dream of expanding the domain of life from earth into the universe makes sense as a long-term goal but not as a short-term goal. The practical feasibility of cheap human voyages and settlement of the solar system depends on fundamental advances in biology. Any affordable program of manned exploration must be centered on biology and will have a timescale tied to the timescale of biotechnology. A timescale of a hundred years is probably reasonable. This is roughly the time it will take us to learn how to grow warm-blooded plants.

When we come to settle the solar system with manned missions at the end of the twenty-first century, we will already have explored the terrain thoroughly with unmanned missions. The people who decide to go to Mars or Europa will know why they decided to go there and what to expect when they get there. They will know whether or not indigenous life exists there. If there is indigenous life, they will know how to nurture and protect it when they come to build their own habitat. If there is no indigenous life, they will bring new life to make good nature's lack. The most important part of their luggage will be the seeds of plants and animals genetically engineered to be capable of survival in an alien climate. On a world that has only a thin atmosphere like Mars, or no atmosphere at all like Europa, the most useful seeds will be the seeds of warm-blooded plants. After a hundred years of development of the art of genetic engineering, we will know how to write the DNA to make plants grow greenhouses. Plants as large as trees could grow greenhouses big enough for humans to live in. If the human settlers are wise, they will arrive to move into homes already prepared for them by an ecology of warm-blooded plants and animals introduced by earlier unmanned missions.

Warm-blooded plants will not by themselves solve all our problems. They are necessary but not sufficient. They are only the first of the thousands of diverse new species that will be required to create viable ecologies in the places where humans may wish to go. A biological technology mature enough to create warm-blooded plants will also be able to take care of other ecological problems, on Mars or on Europa or even at home on Earth. The essential requirement for a successful human colony will be a deep understanding of the local ecology, so that humans can become a part of the ecology without destroying it. The Biosphere 2 experiment in Arizona, where eight humans tried unsuccessfully to live in a closed ecology for two years, was not a failure but a valuable object lesson. It taught us how humans without sufficient understanding of their habitat could unexpectedly run out of oxygen to breathe.

Why should anybody wish to live on Mars or Europa? The only answer we can give to this question is the answer George Mallory gave to the question why he wanted to climb Everest: "Because it is there." There may be economic reasons, scientific reasons, or sentimental reasons attracting people to remote places. People always have a variety of reasons for moving from one place to another. One of the few constant factors in human history is the fact that people migrate, often covering huge distances for reasons that are difficult to discern. As soon as emigration from earth becomes cheap enough for ordinary people to afford, people will emigrate. To make human space travel cheap, we shall need advanced biotechnology in addition to advanced propulsion systems. And we shall need a large number of travelers to bring down the cost of a ticket. That is why human space travel will not be cheap until fifty or a hundred years have gone by.

When we look ahead further than a hundred years, the most important fact about the geography of the solar system is that the habitable surface area is almost all on small objects—asteroids and comets—rather than on planets. Planets have most of the mass but very little of the surface area. The asteroids are objects made mostly of rock and orbiting in the inner part of the solar system, closer to the sun than Jupiter. The comets are objects mostly made of ice and orbiting in the outer part of the solar system, further from the sun than Neptune. Comets of average size are only visible from earth on the rare occasions when gravitational perturbations cause them to fall close to the sun and their volatile surfaces boil off to form bright tails in the sky. A huge swarm of them remains in a ring-shaped region called the Kuiper Belt outside the orbit of Neptune. Only in the last few years have a few of the largest Kuiper Belt objects been seen in their native habitat, first with telescopes on the ground in Hawaii and later with the Hubble telescope in space. A rough estimate indicates that the total surface area of the trillions of objects in the Kuiper Belt is about a thousand times the area of the earth.

There are three reasons why comets are more significant than asteroids in the ecology of the solar system. First, the comets are vastly more numerous. Second, ice is better than rock as a basis for life, and the comets contain not only ice but most of the other chemical elements that are essential for biology. Third, the orbital speeds of the comets are much slower than the speeds of asteroids.

The Kuiper Belt may seem to us today to be a cold and inhospitable place, but it is probably less inhospitable to life than Mars. It has the advantage of being an archipelago, a collection of small habitable islands not too far from one another. Because the relative speeds are slow, communication and travel between one island and another will be easy. When you are living on a billion-ton comet in the Kuiper Belt, another billion-ton object will on the average pass by within a million miles about once a month. Million-ton objects will come by within this distance every day. It will take only a few days, using a small spacecraft with a modest propulsion system, to hop over and visit the neighbors or replenish supplies. If you are bored by the scenery or unhappy with your family, you can move permanently and try your luck on another comet, just as the Pilgrims in the old days moved from Plymouth to Boston and from Boston to Providence and places west.

If a community occupying a Kuiper Belt object

outgrows its habitat and wishes to expand, it can increase its living space by attaching tethers to neighboring objects as they float by. A metropolis could grow by accretion of objects in the twenty-second century as rapidly as Chicago or San Francisco grew by accretion of real estate in the nineteenth. A Kuiper Belt metropolis would probably be a flat, disk-shaped collection of cometary objects, linked by long tethers and revolving slowly around its center to keep the tethers taut. To continue the accretion of desirable properties while avoiding destructive impacts, the metropolitan border patrol would be engaged in an interesting game of celestial billiards, tracking approaching objects with telescopes, nudging them gently with space tugs and hooking them with tethers.

Recently the inhabitants of Earth have become aware that our planet is exposed to occasional impacts of asteroids and comets that may cause world-wide devastation. The most famous such impact occurred sixty-five million years ago in Mexico and may have been responsible for the demise of the dinosaurs. During the next hundred years, as the technologies of astronomical surveillance and space propulsion move forward, it is likely that active intervention to protect the Earth from future impacts will become feasible. We may see a mutually profitable merger of the space science enterprise with the business of planet protection. The cost of protection is modest provided that the warning time before impacts is as long as a hundred years. To deflect an orbiting object enough to cause it to miss the Earth, a slow steady push is much more effective than a nuclear explosion. With a hundred-year warning time, the power required for the steady push is only about two kilowatts for an average-size comet with a mass of a billion tons. Two kilowatts is power on a human scale, not on an astronomical scale. Even as far from the sun as the Kuiper Belt, there is enough power in sunlight to supply two kilowatts with a solar collector of reasonable size. Once human communities are established in the Kuiper Belt, their border patrol authorities will be in a position to offer their services to planet Earth, to detect objects that threaten to collide with Earth and deflect them in timely fashion at minimal cost.

It could well happen within a few hundred years that most of the inhabitants of the solar system will be living in the Kuiper Belt. Accustomed as we are to living on a high-gravity planet close to the sun, it is difficult for us to imagine what it will be like to live in low gravity far away. One of the first steps that a human colony would take to establish itself in the Kuiper Belt would be to surround its cometary habitat with an extended efflorescence of mirrors in space to collect sunlight. An array of mirrors a hundred miles in diameter would collect a steady thousand megawatts of energy anywhere in the Kuiper Belt, out to four times the distance of Neptune from the sun. That is enough energy to sustain a considerable population of plants, animals, and humans with all modern conveniences.

The mirrors would not need to be optically perfect. The material out of which to construct them, a few thousand tons of metal or plastic, will probably be available on any Kuiper Belt object. After a century of progress in biotechnology, we will not need to manufacture the mirrors. We will teach our plants to grow them. Life in the Kuiper Belt will be different from life on Earth, but will not necessarily be less beautiful or more confined. After a century or two, there will be metropolitan centers and cultural monuments and urban sprawl, all the glories and discontents of a high civilization. There will soon be restless spirits who find that the Kuiper Belt is growing too crowded. But there will still be an open frontier and a vast wilderness beyond. Beyond the Kuiper Belt lies another far more extended swarm of comets, the Oort Cloud, further away from the sun and still untamed.

*　*　*

The most important surprises of the next fifty years are likely to come from the internet and the genome rather than from the sun and the sky. Two big surprises happened in 1997, soon after I had given the lectures that formed the basis for this book. The first surprise was the announcement by Ian Wilmut that he had cloned a cell from an adult sheep and thereby given birth to Dolly. The second was the defeat of world chess champion Gary Kasparov by the IBM chess-playing software program Deep Blue. Both these events came as shocks to the general public. The events were unimportant in themselves but were harbingers of future revolutions. The victory of Deep Blue stands as a symbol of the inexorable growth of artificial intelligence in the coming century, and the cloning of Dolly stands as a symbol of the inexorable growth of genetic manipulation. Artificial intelligence and genetic manipulation are indeed long-term threats to the autonomy of the human spirit.

The cloning of Dolly came as a surprise, not only to the general public but also to experts in animal genetics. The experts have been working for many years to understand the processes that convert an unspecialized embryonic cell into a specialized adult tissue cell. The adult cell contains the same genes as the embryonic cell, but most of the genes are switched off in the adult cell so that it does only the specialized job that it is supposed to do. To clone a sheep from an adult cell, Ian Wilmut had to find a way to reverse all the switches so that the adult cell was transformed back into an embryo.

The surprise was the fact that he was able to reverse the switches without understanding them in detail. Ian

Wilmut succeeded because he belongs to the culture of practical animal breeding rather than the culture of analytical genetics. For thousands of years, animal breeders have successfully created new varieties of animals, from dairy cattle to toy poodles, without understanding the details of their genomes. Wilmut worked by trial and error, in the tradition of the old craft industry of animal breeders. He believed there was a good chance that some simple physical and chemical manipulations would reverse the switches, and it turned out that he was right. If he had waited until he understood the molecular structure of the switches before doing the manipulations, he might have produced some interesting science but he would not have produced Dolly. Even in the age of the genome, animal breeding remains halfway between an art and a science.

The research that led to the cloning of Dolly was driven by practical needs. The technology of cloning is not yet efficient and reliable. Dolly was a single success among several hundred failures. Before long, the causes of failure will be understood and the technology will be ready for practical use. It will be used to breed animals with precisely known and guaranteed genetic qualities. Champion cows, sheep, and pigs will be standardized and mass-produced.

But Ian Wilmut also has in mind applications of cloning to the production of drugs to treat human ailments. Many important human diseases are caused by deficiencies of certain proteins that are made in small quantities by healthy people. To extract these proteins in sufficient quantity from human blood or corpses is impossible, but it is possible to transfer a human gene to a goat in such a way that the goat will manufacture a human protein and secrete useful quantities of the protein into her milk. To extract the protein from the milk is not difficult, and the technology for transferring human genes into animal egg cells already exists. At present the gene transfer procedure is inefficient and costly, so that the human proteins are in short supply and are still not available to most of the people who need them. If a single goat carrying a human gene can be cloned, the procedure of gene transfer need only be done once, and the human proteins can thereafter be produced in adequate quantity as needed. Wilmut mentioned human blood-clotting factor and fibrinogen for healing wounds as badly needed products that a cloned goat might be able to supply. He also has a longer-range aim of developing the technology to clone animals whose tissues carry human recognition signals. Tissues from these animals could then be transplanted into humans without causing the human immune system to reject them. This would allow humans to receive organ transplants when they are needed, instead of waiting for years until a compatible human donor happens to die.

When the existence of Dolly was announced and Ian Wilmut suddenly became famous, he was asked whether he intended to clone human beings. His answer was firmly negative. He said he could imagine no good reason to clone a person. He was anxious to continue his work, learning how to clone animals for useful purposes. He wanted to talk about animals and not about humans. He was determined not to become entangled with the thorny moral and political problems that arise as soon as one talks about cloning humans. But the question of cloning humans will not disappear. As soon as the technology of cloning becomes safe and reliable, it will also be available for humans to use on themselves.

Lee Silver, a professor of biology at Princeton University, has studied what has been happening in fertility clinics during the twenty years since Louise Brown, the first human test-tube baby, was born in England. He has written a book, *Remaking Eden*, describing what he found and where it is likely to lead. The facts that he reports are unambiguous. Fertility clinics are the most rapidly growing and the most profitable branch of medicine, not only in economically advanced countries but in many poorer countries all over the world. Everywhere there is a huge demand for babies from parents who for one reason or another cannot produce them in the traditional way. Parents spend large sums of money on procedures that offer only a small chance of a successful pregnancy. From these facts Silver deduces some simple generalizations. The desire to have children has been built into human nature as our species evolved. Nothing we can do will stand in the way of this desire. Parents will stop at nothing to obtain a baby they can call their own.

The technology available in fertility clinics until now is still, with minor variations, the same technology of in vitro fertilization that was used to conceive Louise Brown. The variations are required when the parents are not able to provide their own egg and sperm for fertilization, so that egg or sperm or both must be obtained from other people. The clinics offer a wide range of alternative methods to deal with such cases, using donors who may be male or female, known to the parents or anonymous. These methods are accepted by the parents as the next best thing to having a baby that carries their own genes.

So much for the past and the present. Lee Silver then turns to the future. Before long, two new technologies will be available to parents who come to fertility clinics: first, cloning, and second, deleting and inserting genes in the fertilized egg before it is implanted in the mother's womb. The second of these technologies he calls *reprogenetics*. Reprogenetics would offer to the parent the opportunity to improve the quality of life of the child, to remove genes that carry susceptibility to various diseases, to insert genes that carry real or perceived advantages. In the long run, reprogenetics is far more important than cloning. Cloning would probably be attractive to a small minority of parents who would otherwise need alien donors and prefer to give a child their own genes. Reprogenetics might be attractive to a far larger population of parents, including many who do not suffer from infertility and would not otherwise think of going to a fertility clinic. The cloning of Dolly caused an explosion of public statements calling for laws prohibiting the cloning of humans. But the social consequences of allowing parents to clone babies are unlikely to be serious, provided that the procedures are certified to be safe and do not damage the offspring. Meanwhile, the possible repercussions of reprogenetics are far more severe and have received little public attention.

Lee Silver observes that, after the desire to have children, the desire to have successful children is the next strongest force driving parental behavior. Many parents spend the greater part of their disposable income to send their children to private schools and universities, believing that education is the key to success. If, for a small fraction of the cost of higher education, parents could endow their children with superior genes, the demand for reprogenetics might become irresistible. Superior genes might give a child the ability to be an Olympic figure-skating champion, or to be chief executive officer of a company, or whatever else the ambitious parents might wish. Unfortunately, it would not be possible for the parents to obtain the informed consent of the child before undertaking the experiment. Nobody yet knows whether superior genes exist or how they might be identified. But the progress of knowledge of human genetics is rapid, and the technology of reprogenetics will not be far behind.

Silver believes that, as far into the future as he can see, the technology of reprogenetics will be expensive.

He sees a possible long-range consequence of reprogenetics to be the splitting of humanity into two species
which he calls GenRich and Naturals. GenRich are
people who can afford to give their children genetic
enrichment. Naturals are those who are left behind.
After a few centuries, the division between rich and
poor might become hereditary, and the two classes
might cease to interbreed. All economic and political
power would belong to the GenRich. Naturals could
not compete in the market for influential jobs and
would have to be content with the station in life to
which their genes have called them. Human society
would once again be divided into masters and slaves.
Silver warns us that this nightmare could become reality, if a free market in reprogenetic technology were
carried to its logical conclusion.

I am imagining a different view of the future. I consider it unlikely that reprogenetic technology will
remain permanently expensive. Most of our socially
important technologies, such as telephones, automobiles, television, and computers, began as expensive
toys for the rich and afterwards became cheap enough
for ordinary people to afford. It usually takes about fifty
years for a new technology to become generally affordable. Television took less than fifty years; computers
may take a little longer. I do not see any reason why
reprogenetics should depart from this pattern. At first,
reprogenetics will be expensive for the same reasons
that in vitro fertilization is expensive today. It is done
by specialist medical doctors and highly trained technicians, it requires long hours of their personal attention, and the demand for their services exceeds the
supply. After fifty years the situation is likely to be
different. Procedures will become standardized and
largely automated. Rich patients will still pay more for
personal attention and privacy, but clinics for poor
patients will provide standard treatments at reasonable

prices. It is possible that clinics will supply take-home do-it-yourself kits for parents who are willing to learn the necessary skills. Having reprogenetic babies at home might become a popular hobby, like desk-top publishing today.

Silver expects that an effectively free market in reprogenetic technology will exist, at least in the United States, and that efforts to prohibit or restrict its use by parents will not be successful. He expects the technology to remain in private hands. American traditions encourage the replacement of public services by private institutions. In the United States, private universities have more prestige and more resources than public universities, and the idea of a national health service is strongly resisted. But in many other countries, higher education and health services are provided by the state, and traditions favor the maintainance of high-quality public services.

When reprogenetic technology becomes available, it is likely that many countries will decide to make it available to citizens as a public service. Where it exists as a public service, it may also be available in a deluxe private version to the rich, but the private suppliers will not enjoy a monopoly, and the rich patients will not have it all to themselves. In countries where everyone is legally entitled to reprogenetic technology, a division of society into GenRich and Naturals is unlikely. Even in the United States, if the beginnings of such a division become clearly visible, public outrage could put an end to it, just as public outrage put an end to slavery. In the eighteenth and nineteenth centuries, slavery was defended by believers in the free market. Abolitionists decided that the free market should not extend to human bodies, and their view prevailed. I hope and believe that our descendants will decide, in the fullness of time, that the free market should not extend to human genes.

Silver and I are in agreement when we consider the more remote future. We agree that the technology of reprogenetics will ultimately split humanity into many species, and that the division will not be only between rich and poor. The division will be between different philosophies of life and different ways of living. When desires for different ways of living can be translated into reality, the diversity of desires will be translated into a diversity of species. The technology of reprogenetics will then be only a faster version of the natural process of speciation which has been generating new species ever since life began. Speciation happens naturally when a small population becomes genetically isolated from its neighbors, so that it can adapt independently to changed conditions and find its own ecological niche. Speciation has been, for billions of years, the main engine driving evolution. Established species evolve very slowly, if at all. Only by speciation can evolution move fast. The technology of reprogenetics will be a variation played by humans on nature's theme, allowing evolution to move faster by a creative use of genetic isolation.

The biologist Ursula Goodenough describes, in a recent paper with the title "Rapid Evolution of Sex-related Genes," the genetic machinery by which creatures from algae to yeasts and insects and mammals are enabled to speciate rapidly. Rapid speciation enabled all these creatures to evolve faster than their competitors, and therefore the machinery for rapid speciation is built into their genomes. Rapid speciation requires that the genes involved with mating and sexual recognition mutate more rapidly than other genes. Goodenough finds that the genomes of higher organisms contain just two classes of genes that are programmed for rapid mutation: the genes for the immune system and the genes for sexual mating systems. The immune system must mutate rapidly to respond to invading bacteria and viruses. The sexual mating genes must mutate rapidly to raise genetic barriers between populations. Genetic barriers are nature's way of creating new species. When we have mastered the technology of reprogenetics, we shall be creating our own genetic barriers, not in opposition to nature, but enabling the natural processes of human evolution to continue.

Even if the division of humanity into several species is a division among equals and not a division between masters and slaves, it will still bring with it intractable social and ethical problems. It is difficult to imagine several human species coexisting peacefully on this small and crowded planet. Here Silver and I are again in agreement. To allow the diversification of human genomes and lifestyles on this planet to continue without restraint is a recipe for disaster. Sooner or later, the tensions between diverging ways of life must be relieved by emigration, some of us finding new places to live away from the Earth while others stay behind. In the end we must travel the high road into space, to find new worlds to match our new capabilities. To give us room to explore the varieties of mind and body into which our genome can evolve, one planet is not enough.

epilogue

The defeat of human chess champion Gary Kasparov by the software program Deep Blue in May 1997 did not come as a surprise to people who were familiar with the pace of computer development. Computers and software evolve ten thousand times more rapidly than humans. Sooner or later, a chess-playing software package was bound to beat a human champion. Still, we must grieve with Kasparov for his failure. A human grandmaster is an artist, creating patterns of movement on a chessboard as a painter creates patterns of color on a canvas. The art of a great chess player is as mysterious as the art of a great painter. The defeat of an artist by a machine is a genuine tragedy. It was rightly seen by the public as a historic event, a symbol of the increasing dominance of machines over human judgment in every corner of our lives.

The suddenness of Kasparov's defeat made it more damaging to human pride. Kasparov went into the battle overconfident and came out exhausted and depressed. He had expected a quick victory and was overwhelmed by his quick defeat. The conditions of the contest ought to have been arranged like the historic contest between Bobby Fischer and Boris Spassky in 1972, a series of twenty-one games, with ample time for the two contestants to get to know each other's strengths and weaknesses, to recover from temporary setbacks, to learn from their mistakes. The 1972 contest was a magnificent drama, with temperamental outbursts in the first act, intellectual fireworks in the second act, peaceful closing in the third act. At the end came Spassky's gracious admission of defeat. Brash brilliance had defeated cool classicism. This was a fair contest in which both sides had a chance to do their best. In contrast, the battle between Kasparov and Deep Blue in 1997 was a hurried affair, with a total of six games.

Kasparov was demoralized by a single lost game and had no time to recover. Deep Blue had access to a database of games played by Kasparov over many years, while Kasparov had access only to the few games that Deep Blue played against him. We may hope that the next match between human champion and machine may be played like the 1972 match, with equal access of both sides to the other's games, with enough time for the human to study the style of the machine and to play against it with some degree of understanding. Then we might see a contest of styles as dramatic as the contest of 1972. If the machine wins again, as it probably will, both sides will have gained from the contest new insights into the game's mysteries.

The victory of the machine will not mean the end of chess as an art form. In the long run, the art of chess will be enriched by the successors of Deep Blue. There will be three separate types of tournament: some for humans only, some for machines only, and some for humans aided by machines. All three types will give scope for artistry and for deeper understanding of the game. The third type will bring the deepest combination of artistry with understanding. The symbiosis of human champion with analytical machine will carry chess far beyond the point where Fischer and Kasparov stopped.

What are the implications of Deep Blue's victory for human society as a whole? Chess is a highly artificial pursuit, of no more direct relevance to the majority of humans than astrophysics or speleology. But there is a strong and valid analogy between the impact of Deep Blue on chess players and the impact of computer networks on ordinary people. For human society as a whole, an individual machine or an individual program like Deep Blue poses no threat. The threat to the dignity of humans and to the autonomy of our institutions comes from the proliferation of little machines in our homes and offices, joined together by inscrutable networks made of telephone cable and optical fiber. The little machines are turning our five-year-old grandchildren into computer addicts and turning our business managers into computer interfaces. The internet and the World Wide Web are permeating our society and changing the way we live. The average citizen of the world, who lacks specialized training and knowledge, can neither escape nor control the rampant growth of the networks. The network packet-switching protocols, which were originally designed to operate a command-and-control system robust enough to survive in the chaos of thermonuclear war, are admirably suited to thrive in the chaos of modern mass communications. The networks of today are embryonic forms, destined to grow into mature structures whose shape and power we cannot yet imagine. The defeat of Kasparov is a metaphor for the human condition as we shall be, if we let ourselves be blindsided by the growth of networks.

On the other hand, the more benign future of chess that I envisage, with human artistry and computer power evolving together in a creative symbiosis, is also a metaphor for a possible future evolution of human society in a world of networks. Evolution in the past has always been driven by a shifting balance between competition and symbiosis. So it must be in the future. It is our task as humans to keep the balance in equilibrium. The balance today is out of control and tilting sharply.

The networks are driving us into a world of cut-throat competition that many of us find destructive. The networks impose cultural and economic constraints that we feel powerless to resist. The networks mostly serve the rich and are inaccessible to the poor and uneducated, thereby increasing the barriers and inequalities between rich and poor. To this injury they add insult, threatening to reduce humans to the status of cells in a multicellular organism that is indifferent to our needs and desires. But we have the power as individuals to make our needs and desires heard. As creators of the machines and protocols by which the networks live, we have the power to understand them and to influence their functioning. We have the responsibility for making the networks serve the interests of social justice and human freedom. The game of evolution, like the game of chess, will in future be played by humans and machines working together. The landscape of cyberspace offers us as much scope for artistic creation as the landscape of a chessboard.

References

This list of references makes no claim to be complete. I have included only books and papers that I found helpful. They are listed in the order of their appearance in the text.

Introduction

Godfrey H. Hardy, *A Mathematician's Apology* (Cambridge: Cambridge University Press, 1940), p. 60.

Jean Jules-Verne, *Jules Verne: A Biography*, trans. Roger Greaves (New York: Taplinger Publishing Co., 1976). See p. 190 for "An Ideal City," a talk that Verne gave at Amiens, a rather sleepy provincial town, making fun of the ambition of some of the citizens to develop the town into a modern city.

H. G. Wells, *When the Sleeper Wakes*, in *Three Prophetic Novels of H. G. Wells*, selected and with an introduction by E. F. Bleiler (New York: Dover Publications, 1960). The first edition of *When the Sleeper Wakes* was published by Harper in 1899.

Edward Tenner, *Why Things Bite Back: Technology and the Revenge of Unintended Consequences* (New York: Alfred A. Knopf, 1996). The reference to Albert Robida is on p. 17.

Esther Dyson, *Release 2.0, A Design for Living in the Digital Age* (New York: Broadway Books, 1997).

George B. Dyson, *Darwin Among the Machines: The Evolution of Global Intelligence* (New York: Addison-Wesley, 1997).

Chapter 1. Scientific Revolutions

Part of Chapter 1 was published with the title "Science as a Craft Industry" in *Science* 280 (1998): 1014-15.

Henry A. H. Boot and John T. Randall, "Historical Notes on the Cavity Magnetron," *IEEE Transactions on Electron Devices* 23 (1976): 724-29. See also Robert Buderi, *The Invention that Changed the World* (New York: Simon and Schuster, 1996), especially ch. 4, pp. 82-89.

Tracy Kidder, *The Soul of a New Machine* (Boston: Little Brown and Company, 1981).

Peter Galison, *Image and Logic: A Material Culture of Microphysics* (Chicago: University of Chicago Press, 1997). For Marietta Blau, see pp. 148-60.

Thomas Kuhn, *The Structure of Scientific Revolutions*, 2nd ed. (Chicago: University of Chicago Press, 1970). The first edition was published in 1962.

Alan D. Sokal, "Transgressing the Boundaries: Towards a Transformative Hermeneutics of Quantum Gravity," *Social Text* 14 (1996): 217-52. To be fair, anyone who reads Sokal's parody should also read Andrew Ross, "Reflections on the Sokal Affair," *Social Text* 15 (1997): 149-52, a thoughtful response by one of the editors of *Social Text*.

Emily Martin, "Understanding the Immune System Culturally" (lecture given at the Institute for Advanced Study, Princeton, N.J., February 1993).

Alexander Wolszczan and D. A. Frail, "A Planetary System around the Millisecond Pulsar PSR 1257+12," *Nature* 355 (1992): 145-47. Alexander Wolszczan, "Confirmation of Earth-Mass Planets Orbiting the Millisecond Pulsar PSR B1257+12," *Science* 264 (1994): 538-42. For the demolition of Wolszczan's claim to have observed a third planet with a period of 25.34 days, see Klaus Scherer et al., "A Pulsar, the

Heliosphere, and Pioneer 10: Probable Mimicking of a Planet of PSR 1257+12 by Solar Rotation," *Science* 278 (1997): 1919-21.

John A. Sidles, Joseph Garbini, and Gary Drobny, "The Theory of Oscillator-coupled Magnetic Resonance with Potential Applications to Molecular Imaging," *Review of Scientific Instruments* 63, no. 8 (1992): 3881-99.

Bogdan Paczyński, "Gravitational Microlensing in the Local Group," *Annual Review of Astronomy and Astrophysics* 34 (1996): 419-59.

Chapter 2. Technology and Social Justice

This chapter is an expanded version of the second New York Public Library lecture. The expanded version, with minor changes, was presented as a Louis Nizer Lecture to the Carnegie Council on Ethics and International Affairs in New York on November 5, 1997. The text of the lecture was published by the Carnegie Council in March 1998 with the title *Technology and Social Justice*. I am grateful to the Carnegie Council for permission to reprint it here.

Freeman J. Dyson, *Infinite in All Directions* (New York: Harper and Row, 1988), p. 273.

Max Weber, *The Protestant Ethic and the Spirit of Capitalism*, trans. Talcott Parsons with a foreword by R. H. Tawney (London: George Allen and Unwin, 1930), (Tübingen: Mohr, 1923). R. H. Tawney published his own account of the history in *Religion and the Rise of Capitalism* (New York: Harcourt Brace, 1926). A contrasting view is provided by Michael Novak, *The Catholic Ethic and the Spirit of Capitalism* (New York: Free Press, 1993). See also Gertrude Himmelfarb, *The De-moralization of Society* (New York: Knopf, 1995), ch. 6, for a discussion of the relation between the Jewish ethic and the rise of capitalism. It seems that Protestants, Catholics, and Jews can all make valid claims to have assisted at the birth of modern Europe.

The headquarters of Solar Electric Light Fund is at 1734 20th Street NW, Washington, DC, 20009. The E-mail address is solarlight@self.org.

For details of Paul Baran's Ricochet system, see George B. Dyson, *Darwin Among the Machines*, pp. 207-8.

After I gave this lecture at the University of California, Irvine, in May 1998, Virginia Trimble, a professor of astronomy, asked the audience for a show of hands. "How many of you had grandmothers with servants in the home?" About fifty hands went up. Then she asked for another show of hands. "How many had grandmothers who were themselves servants?" Two hands went up. If one can believe these numbers, the audience was mainly descended from wealthy families. Upward mobility was not frequent. Since the audience consisted of professional academics, it appears that the academic profession, at least in California, is in danger of becoming a hereditary caste.

Chapter 3. The High Road

Part of this chapter was published with the title "Warm-blooded Plants and Freeze-dried Fish" in *The Atlantic Monthly* 280 (November 1997): 71-80.

For the history of the Mars rock ALH84001, see *Planetary Report* 17 (January 1997). This issue contains articles by Roberta Score, who found it, and by David Mittelfehldt, who identified it. The evidence for traces of ancient life in the rock was presented by David McKay, et al., "Search for Past Life on Mars: Possible Relic Biogenic Activity in Martian Meteorite ALH

84001," *Science* 273 (1996): 924-30. For high-resolution pictures of the ice on Europa, see *Sky and Telescope* 93 (March 1997): 32-33.

For the evidence that the interior of the Mars rock never experienced high temperatures, see D. A. Evans, N. J. Beukes, and J. L. Kirschvink, "Paleomagnetic Evidence for a Low-temperature Origin of Carbonate in the Martian Meteorite ALH84001," *Science* 275 (1997): 1629-33.

For terrestrial warm-blooded plants, see Roger S. Seymour, "Plants that Warm Themselves," *Scientific American* 276 (March 1997): 104-9 .

Leik Myrabo, "Laser Propelled Flight Experiments at the High Energy Laser Systems Test Facility," in *Space Manufacturing 11,* Proceedings of the Thirteenth SSI/Princeton Conference on Space Manufacturing, ed. Barbara Faughnan (Princeton, Space Studies Institute, 1997), pp. 142-43.

C. Knowlen, A. P. Bruckner, D. W. Bogdanoff, and A. Hertzberg, "Performance Capabilities of the Ram Accelerator," in Proceedings of the American Institute of Aeronautics and Astronautics Twenty-third Propulsion Conference, San Diego, June-July 1987. A. P. Bruckner and A. Hertzberg, "Ram Accelerator Direct Launch System for Space Cargo," in Proceedings of International Astronautical Federation Thirty-eighth Congress, Brighton, UK, October 1987.

Derek Tidman, "Slingatron Dynamics and Launch to Low Earth Orbit," in *Space Manufacturing 11, op. cit.*, pp. 139-41.

I. Wilmut, et al., "Viable Offspring Derived from Fetal and Adult Mammalian Cells," *Nature* 385 (1997): 810-13.

Lee M. Silver, *Remaking Eden* (New York: Avon Books, 1997). For the essential facts on which Silver bases his argument, see ch. 6, pp. 67-77.

Ursula W. Goodenough, *Rapid Evolution of Sex-related Genes*, private communication (1998).

Epilogue

The epilogue was published with the title "Chess and the Human Condition" in *Think*, the magazine published by IBM for its employees, Fall 1997.

A NOTE ON THE TYPE

The Sun, the Genome, & the Internet is set in Adobe Caslon and Democratica. Carol Twombly drew Caslon in 1989, basing her work on that of the eighteenth-century English engraver William Caslon. Miles Newlyn drew Democratica, the display font, in 1991.

Design and composition by Adam B. Bohannon